Integrated Cattle Nutrition

Integrated Cattle Nutrition

Balaji Yadav Maddina

RANDOM
RANDOM PUBLICATIONS
NEW DELHI (INDIA)

Integrated Cattle Nutrition

ISBN 978-93-5111-953-1

Published in 2016 in India by

RANDOM PUBLICATIONS

4376-A/4B, Gali Murari Lal, Ansari Road
New Delhi-110 002
Phone : +9111-43580356, 011-23289044, 011-43142548
e-mail: sales@randompublications.com,
info@randompublications.com, randomexports@gmail.com

Type Setting by : Friends Media, Delhi-110089
Digitally Printed at : Replika Press Pvt. Ltd.

Preface

Animals need a variety of nutrients to meet their basic needs. These nutrients include fats and carbohydrates that provide energy, proteins that furnish amino acids, vitamins that serve as co-factors for enzymes and perform other functions, ions required for water balance and for nerve and muscle function, and selected elements that are incorporated into certain molecules synthesized by cells. To determine the levels of nutrients that are needed to sustain normal activities, researchers monitor the relationship between nutrient intake, the levels of nutrients maintained in the body, and health. To determine how nutrition can affect athletic performance, researchers alter the intake of specific nutrients and assess how athletes respond. For example, investigators have shown that athletes who ingest a high-carbohydrate diet for three days outperform individuals who ingest a high-fat, high-protein diet.

Feeding diets with lowered protein content reduces nitrogen input, improves nitrogen utilization efficiency, and reduces nitrogen losses from manure. Reducing dietary protein also benefits the producer by reducing feed cost and improving overall farm profitability. These interventions, however, have to be balanced with the risk of loss in milk production. If the true animal requirements for metabolizable protein are not met, long-term production cannot be sustained.

Cells use nutrients as fuel for energy production (catabolism) and as material for processes of maintenance and growth (anabolism). Multicellular animals derive energy solely from the breakdown of complex organic molecules, mainly carbohydrates and fats. Because the fuel for the maintenance of animal life comes only from other living organisms or their remains, animals are known as heterotrophic organisms. All animal life depends ultimately on the existence of organisms (largely green plants) that can use inorganic sources of energy, of which solar radiation is by far the most important; some microorganisms, however, obtain energy from oxidation of simple inorganic compounds.

The contents besieged latest techniques on cattle nutrition, animal nutrition, rumen microbes enumeration, isolation and characterization with economized methods of feed evaluation.

– ***Author***

Contents

1

Animal Nutrition

Animals need a variety of nutrients to meet their basic needs. These nutrients include fats and carbohydrates that provide energy, proteins that furnish amino acids, vitamins that serve as co-factors for enzymes and perform other functions, ions required for water balance and for nerve and muscle function, and selected elements that are incorporated into certain molecules synthesized by cells. To determine the levels of nutrients that are needed to sustain normal activities, researchers monitor the relationship between nutrient intake, the levels of nutrients maintained in the body, and health. To determine how nutrition can affect athletic performance, researchers alter the intake of specific nutrients and assess how athletes respond. For example, investigators have shown that athletes who ingest a high-carbohydrate diet for three days outperform individuals who ingest a high-fat, high-protein diet.

There are seven major classes of nutrients: carbohydrates, fats, fibre, minerals, protein, vitamin, and water. These nutrient classes can be categorized as either macronutrients (needed in relatively large amounts) or micronutrients (needed in smaller quantities). The macronutrients are carbohydrates, fats, fibre, proteins, and water. The micronutrients are minerals and vitamins.

The macronutrients (excluding fibre and water) provide structural material (amino acids from which proteins are built, and lipids from which cell membranes and some signaling molecules are built) and energy. Some of the structural material can be used to generate energy internally, and in either case it is measured in joules or calories (sometimes called "kilocalories" and on other rare occasions written with a capital C to distinguish them from little 'c' calories). Carbohydrates and proteins provide 17 kJ approximately (4 kcal) of energy per gram, while fats provide 37 kJ (9 kcal) per gram.,[1] though the net energy from either depends on such factors as absorption and digestive effort, which vary substantially from instance to instance. Vitamins, minerals, fibre, and water do not provide energy, but are required for other reasons. A third class dietary material, fibre (*i.e.*, non-digestible material such as cellulose), seems also to be required, for both mechanical and biochemical reasons, though the exact reasons remain unclear.

Molecules of carbohydrates and fats consist of carbon, hydrogen, and oxygen atoms. Carbohydrates range from simple monosaccharides (glucose, fructose, galactose) to complex polysaccharides (starch). Fats are triglycerides, made of assorted fatty acid monomers bound to glycerol backbone. Some fatty acids, but not all, are essential in the diet: they cannot be synthesized in the body. Protein molecules contain nitrogen atoms in addition to carbon, oxygen, and hydrogen. The fundamental components of protein are nitrogen-containing amino acids, some of which are essential in the sense that humans cannot make them internally. Some of the amino acids are convertible (with the expenditure of energy) to glucose and can be used for energy production just as ordinary glucose. By breaking down existing protein, some glucose can be produced internally; the remaining amino acids are discarded, primarily as urea in urine. This occurs normally only during prolonged starvation.

Other micronutrients include antioxidants and phytochemicals which are said to influence (or protect) some body systems. Their necessity is not as well established as in the case of, for instance, vitamins. Most foods contain a mix of some or all of the nutrient classes, together with other substances such as toxins or various sorts. Some nutrients can be stored internally (*e.g.*, the fat soluble vitamins), while others are required more or less continuously. Poor health can be caused by a lack of required nutrients or, in extreme cases, too much of a required nutrient. For example, both salt and water (both absolutely required) will cause illness or even death in too large amounts.

NUTRIENT-NUTRIENT INTERACTIONS

There are roughly 60 physiologically essential nutrients, and most of those that are present in foods and in the body do not exist as the nutrient *per se* but instead as components of proteins, fats, enzymes, coenzymes, nucleic acids, phytates, or bone components. Work with experimental animals has led to information on the structure of many of these nutrient-containing compounds. Clearly, digestive and (or) metabolic processes are required to transform these compounds into a form that the body can absorb from the gut, metabolize in the body, and use.

Animal models were used to characterize most of the classic mineral interactions. We now know that excess dietary P decreases Ca absorption, and likewise, that excess dietary Ca decreases P absorption. Also, vitamin D is required for efficient absorption of both Ca and P. The interactions between Zn and Cu; Mo and Cu; Zn and Fe; cysteine or ascorbate and Cu, Zn, or Fe; Mn and P; Se and vitamin E; I and Br; As and cysteine; phytates or oxylates and several macro- and micromineral elements have been identified and help explain the utilization efficiencies of mineral elements in various foods. Among the vitamins, animals were used to help us understand the tryptophan × niacin and lysine × niacin interactions as well as the effect of Fe deficiency on the conversion efficiency of tryptophan to niacin. Also, progress in solving pellagra in humans was held back for many years because of lack of a suitable animal model. Ultimately, the dog (black tongue) became the best animal model for learning which foods were pellagragenic and which were ameliorative.

The rat and pig were used to prove that excess dietary methionine (also *S*-methylmethionine) could spare, and even eliminate, the dietary need for preformed choline. The close functional relationship between folic acid and vitamin B-12 was also worked out using animal models, as was the establishment that both riboflavin and Zn were required to convert vitamin B-6 vitamers to pyridoxal phosphate. Before vitamin B-12 became available in crystalline form, clues as to its presence in animal or fermentation products but not plant products came from the "animal protein factor" work done with chickens and pigs in the period 1940 to 1950.

The many amino acid interactions/interrelationships also profited from animal research. Thus, the methionine-cysteine, phenylalanine-tyrosine, arginine-proline and glycine-serine interactions are all examples of 1 amino acid sparing the dietary (or physiological) need for another amino acid. In other cases, negative effects of 1 or more amino acids were shown to affect utilization of other amino acids. This is illustrated with antagonisms among the branched-chain amino acids in rats and chicks, lysine antagonizing arginine in chicks and dogs but not in pigs or cats, and supplemental glycine being capable of partially reversing growth depressions in chicks and rats caused by excess methionine consumption. The negative interactions among amino acids have, for the most part, not been studied in humans.

BIOAVAILABILITY IN EXPERIMENTAL ANIMALS

There are probably no better examples of how experimental animals have contributed basic nutritional information than studies conducted on bioavailability of nutrients and nutrient precursors. For the most part, energy values of foods, digestibility and protein quality of proteins, absorption efficiencies of mineral salts and the minerals in foods, absorption efficiencies of the vitamin components of foods and various vitamin precursors, and amino acid utilization from various amino acid precursors, analogs, and metabolites have been derived from studies employing animal models. Today, however, stable isotope techniques are allowing more and more nutrient bioavailability studies to be done using human subjects.

Absorption efficiency of mineral elements

There are vast differences in the absorption efficiency of mineral elements provided either as mineral salts or as components of food. The halogen elements Cl, I, and F are almost completely absorbed, and Na and K are generally absorbed with over 90 per cent efficiency. In contrast, the essential mineral elements Cr and Mn are absorbed poorly, most estimates being in the range of 2 per cent to 4 per cent. Phosphorus from inorganic sources is absorbed with 70–90 per cent efficiency, depending on P status, but much of the P in plant-based foods exists as a phytate complex, and in this form the efficiency of P absorption may be as low as 10 per cent. The P in foods existing in either phospholipid or nucleic acid form, however, is absorbed as efficiently as inorganic P. Phytic acid in foods also has negative effects on the absorption of other mineral elements such as Ca, Zn, Fe, and Mn. Oxalates in foods such as spinach have similar effects. The negative effects of phytic acid on trace-mineral bioavailability are exacerbated by excessive Ca ingestion. Reducing agents such as cysteine and ascorbate enhance the absorption of Zn and Fe but decrease the absorption of Cu and Co.

Animal studies have led the way in providing answers to questions of bioavailability of mineral supplements ingested as mineral salts. Sulfates, chlorides, phosphates, acetates, citrates, and gluconates are generally the best sources of supplemental elements, but carbonates of Ca and Zn are well utilized also. Oxides and sulfides are generally poor sources of mineral elements, although food-grade ZnO and MnO are well utilized as bioavailable sources of Zn and Mn. Oxides of Fe and Cu have relative bioavailability values (*i.e.*, efficacy value relative to a standard such as $FeSO_4·7H_2O$ or $CuSO_4·5H_2O$) of near zero.

Bioavailability of Vitamins in Foods and Food Ingredients

Bioavailability of vitamins in foods and food ingredients has received little research attention, probably because it is easy and economical to supplement diets with crystalline forms of the vitamins. Nonetheless, in poor countries of the world, crystalline vitamins are often unavailable or unaffordable, such that multiple vitamin deficiencies are not uncommon. Animal models were instrumental in solving vitamin-deficiency diseases such as beri-beri (chick thiamin deficiency), scurvy (guinea pig ascorbic acid deficiency), pellagra (dog, rat, pig, and chick deficiency of niacin and tryptophan), rickets (dog, rat, and chick deficiency of Ca, P, and/or vitamin D), night blindness (rat deficiency of vitamin A), dermatitis (rat deficiency of vitamin B-6), low fertility and muscle dystrophy (rat deficiency of vitamin E), hemorrhagic disease (chick deficiency of vitamin K), and anemia (monkey, rat, and chick deficiency of folate and/or vitamin B-12).

Many of these diseases were thought initially to be infectious and of bacteriologic origin. However, when an association with diet was noted, and when this was followed by development of animal-model bioassays with defined purified diets, progress was quickly made in defining the disease condition and in reversing or preventing it with the proper vitamin-containing food or (later) with the vitamin itself. Today we know that certain vitamins are extremely heat labile (thiamin, folate, pyridoxal forms of B-6) because of Maillard-type reactions. Also, the niacin in most cereal sources is poorly available, biotin is antagonized by avidin in raw egg white, and the biotin in wheat is not well absorbed; vitamin D bioactivity is enhanced by UV light, whereas UV light can inactivate riboflavin that is in solution; (excess) methionine can spare the requirement for preformed choline, raw fish may contain antithiamin (thiaminase) factors, and oilseed products (*e.g.*, soybeans) may contain a substantial portion of their vitamin B-6 in glycosylated forms such as pyridoxine glucoside, a poorly bioavailable source of vitamin B-6.

Among the fat-soluble vitamins, vitamin A and vitamin D have received the most attention. Although preformed vitamin A (esters of all-trans retinol) is found in foods of animal origin, carotenoids are the source of vitamin A activity in plant-based foods. At least 600 different carotenoids exist as yellow, orange, and red pigments in foods, but <10 per cent of these have meaningful vitamin A bioactivity. β-Carotene is the carotenoid with the greatest vitamin A activity, but its cleavage efficiency in the gut by 15,15'-dioxygenase is very different among animal species: the chick and rat are as much as 4 times more efficient than the pig in converting β-carotene to retinol. There are 2 main factors that affect the bioavailability of β-carotene and other carotenoids as vitamin A precursors: 1) release of carotenoids from the food matrix and 2) efficiency of the dioxygenase enzyme in splitting all-trans β-carotene into 2 retinal molecules.

These 2 factors, however, are affected by 3 other factors: 1) species, 2) nutrient status, and 3) intake level of provitamin A carotenoids.

In the human, it is assumed that 12μg all-trans β-carotene in a mixed diet will yield 1 μg of all-trans retinol. Because various animal species absorb carotenoid compounds with greatly different efficiencies, selection of a suitable animal model for human vitamin A studies is problematic. Lee *et al.* concluded that gerbils and preruminant calves were the best models for evaluating β-carotene conversion to vitamin A in humans. Other species, however, are more appropriate for studies of the effects of certain xanthophyll compounds (*e.g.*, lutein, zeaxanthin) on macular degeneration and cataract formation.

Vitamin E deficiency exacerbates vitamin A deficiency in that vitamin E is necessary for cleavage of β-carotene to retinal. Vitamin E exists in foods in 8 different forms, 4 of which are tocopherols and 4 tocotrienols. D-α-Tocopherol has the highest vitamin E bioactivity, but although β-and γ-tocopherols contain substantially lower vitamin E bioactivity values, these isomers of tocopherol nonetheless contain good antioxidant activity. The rat antisterility assay and beef cattle tissue tocopherol uptake assay have been compared with serum vitamin E uptake values in humans. The beef cattle estimates for the bioactivity of D-α-tocopherol, DL-α-tocopheryl, D-α-tocopherol acetate, and DL-α-tocopheryl acetate are more in line with the human estimates than are the rat bioactivity estimates.

There is little information on the bioavailability of vitamin D in food sources vs. that furnished via UV sunlight exposure. However, it is now known that the active form of vitamin D-3 (cholecalciferol) is actually a hormone, 1,25-dihydroxycholecalciferol. This compound is available for clinical use, and it is the most biologically active form of vitamin D. 1 α-Hydroxycholecalciferol, however, is a precursor synthetic form of cholecalciferol that the liver can efficiently convert to the 1,25 hydroxylated form of cholecalciferol. 25-Hydroxycholecalciferol is the principal blood form of cholecalciferol, and this compound, available commercially, has more bioactivity than cholecalciferol itself but lower bioactivity than the 1 α-hydroxylated forms of cholecalciferol.

Virtually nothing is known about vitamin K bioavailability in foods. All 3 forms of vitamin K—phylloquinones (K-1) in plants, menaquinones (K-2) in fermentation products, and menadiones (K-3), *i.e.*, water-soluble synthetic forms of vitamin K—are bioactive. Bioavailability comparisons among the several menadione sources have been done exclusively in animal models, primarily chicks, using prothrombin time as a response criterion.

UPPER TOLERANCE LEVELS FOR NUTRIENTS AND NON-NUTRIENTS

There is a paucity of human information on excess ingestion of both nutrients and non-nutrients. Studies in animals have shown that most nutrient and drug toxicities are exacerbated when purified diets based on casein or amino acids (*e.g.*, enteral and parenteral formulas) are consumed. This is illustrated by work in our laboratory on Cu toxicity in the young chick and methotrexate toxicity in adult rats. The National Research Council has compiled information on both mineral and vitamin tolerances of animals, and these publications are very helpful in summarizing upper tolerance limits for these nutrients. For amino acids, FASEB issued a report in 1992 that dealt with safety issues surrounding excess amino acid intakes, but it was obvious from this report as well as the 2002 Food and Nutrition Board report that most of what is known about amino acid tolerance levels has come from animal research.

The safety margin for excess consumption of vitamins is large, even for vitamin A, which is considered perhaps the most dangerous of the vitamin excesses. Animal studies have predominated over human studies on vitamin excesses, with 2 notable exceptions: niacin and vitamin C. Because pharmacologic niacin doses of 1 to 3 g/d, 67 to 200 times the adult requirement, are used clinically to treat hypercholesterolemia and schizophrenia, the human literature is extensive on side effects (*e.g.*, flushing) of excess niacin ingestion. In terms of growth depression in chicks, dietary levels of niacinamide above 5000 mg/kg cause a linear decrease in weight gain, but niacin levels as high as 20,000 mg/kg (600 mg/d; 3 g·kgBW"^{1}·d"^{1}) are without effect. Regarding vitamin C, with the exception of guinea pigs, apes, and a few birds and fish, animals do not require a dietary source of ascorbic acid. Thus, human studies of vitamin C excess are as prevalent as animal studies. With mineral excesses, there is concern about excess Se and F among the essential mineral elements. Also, there is extensive animal literature on excess consumption of non-nutrient elements such as Cd, Hg, Pb, and As. The chick has been particularly helpful as an animal model for the Cu toxicity problems associated with Wilson's Disease and Indian Childhood Cirrhosis. Supplements ofL-cysteine, L-ascorbate, and Zn are used by hepatologists to treat these Cu toxicity diseases. Chick work has shown that L-cysteine is more effective than either Zn or L-ascorbate in ameliorating Cu toxicity; also, cysteine is more effective than either cystine or methionine. The chick was also used to define I toxicity and its reversal by administration of supplemental Br. Both rats and humans have been used to evaluate pharmacologic doses of Se for cancer prevention.

There is keen interest in the potential benefits of consuming amino acid supplements at levels beyond nutritional requirements. Unfortunately, little evidence exists for tolerable upper amino acid limits in humans. Most of what is known about pharmacologic amino acid dosing comes from studies in animals. It is clear from these studies that individual amino acid excesses are more anorexigenic when provided in low-protein than in high-protein diets. Most of this early work was done with rats fed casein-based purified diets. Use of diets containing mixed ingredients and with normal protein levels is probably more relevant in terms of extrapolation to humans. Our work with chicks and pigs, therefore, involved individual amino acid excesses in diets containing 20 per cent to 23 per cent protein from corn and soybean meal. This work showed that a large dietary excess (4 g/100 g) of methionine, cysteine, or tryptophan was more growth depressing than the same excess of other amino acids. Also, branched-chain amino acids were better tolerated than excesses of other amino acids.

COW DIET

Now what, necessarily, do cows eat? Calves, when first born, are fed milk, much like human infants. Unlike human babies, however, calves can walk within an hour of being born, and three days later they are introduced to calf grain. Whether or not they start eating it is up to them. Calf grain is mainly comprised of corn followed by other grain products, such as wheat and oats. And, to sweeten the deal, molasses is added, giving the grain a sweet smell. As humans grow, so do their appetites, which are the same as those of dairy cows. During the four- to five-week range, calves are introduced to hay and cow feed (as opposed to calf feed).

Hay is dried alfalfa — so in human terms, it's a salad! Then in the six-to eight-week range, dairy farmers will wean calves off milk — that is stop feeding it to them and start pushing more solids and water.

Now the calves become young adults or teenagers, and they're called heifers. Heifers are put together in a group until they become pregnant. Once a heifer has her first calf, she becomes a cow. In order for a cow to continue to produce milk, she has to have a baby every year. Adult cows are like adult humans — they have a job. Theirs is to produce milk. Cows are fed a number of things, including cow feed, hay, grass (in the spring and summer) and silage. Silage can be either corn or alfalfa. Corn silage is slightly different because it contains the whole stalk of corn. Of course, cows receive tons and tons of water, as it takes five gallons of water to produce one gallon of milk. Happy cows equal more milk!

Corn has been a staple ingredient in lactating dairy cow rations for a hundred years. However, many dairy farms import corn to feed to their dairy cows. The price of this imported corn has been fairly stable for a number of years.

Fig. Feeding Dairy Cow

The government is trying to reduce dependence on foreign oil by supporting efforts to produce fuel (ethanol) from renewable resources like corn or cellulose. There is widespread public support for incorporating ethanol into fuelburning vehicles (gasoline and diesel) because ethanol addition results in greater combustion of these fuels and is non-toxic and biodegradable in the water and the soil. The Congress passed legislation called a renewable fuels standard (RFS) that will at least double the use of ethanol and biodiesel by the year 2012. The number of plants producing ethanol from corn is projected to increase dramatically.

This growth in ethanol production has reduced the amount of corn available for livestock feed and concurrently increased its market price. Increased corn prices have resulted in increased feed costs. While higher milk prices may compensate for ration cost increases, partially replacing corn with other less expensive feedstuffs may be required over the longer term.

ALTERNATE RATION STRATEGIES

Other Grain Sources of Dietary Starch

Fig. Grain Diet for cow

Grains, like corn, are an excellent source of starch which is highly fermentable by ruminal microorganisms. The propionic acid produced by starch-fermenting bacteria is converted to glucose by the cow's liver; this glucose is used to make milk. In addition, the bacteria provide about 50 to 60 per cent of the protein needs of the cow as they are washed out of the rumen and are digested in the abomasum and small intestine. To optimize milk production, starch should make up 24 to 26 per cent of the dietary dry matter. Barley and wheat are rich sources of dietary starch. Shifting cow diets between corn and barley or wheat needs to be done slowly.

The starch in barley and wheat is more rapidly available than that in corn so quick dietary changes can result in digestive upsets and lowered ruminal pH. Wheat flour can sometimes be bought for less than corn, but it is very dusty and potentially explosive so it is often avoided by mills. Field reports suggest that flour might 'paste up" in the rumen, so it is not a popular feed ingredient. Hominy contains less starch but more protein, fibre, and fat compared to corn, however their energy density is similar.

Replacing all of the corn with hominy will reduce the dietary starch concentration about 3 percentage units. Dietary costs would be expected to be reduced by replacing corn with hominy. If no milk is lost, then profit is improved. Experiments comparing performance of cows fed ground corn versus hominy could not be located. However, many Florida dairies feed hominy successfully. Even replacing half the corn with hominy can be a reasonable strategy. Because hominy is higher in fat and phosphorus than corn, care should be taken to avoid overfeeding these nutrients.

Low Starch Feedstuffs as Possible Substitutes for Corn

Because grain sources rich in starch, such as barley or wheat, may not be economically available year-round, other commodity feeds must be considered.

Ingredients with lower prices than corn should be considered as partial replacements for corn. Most common feedstuffs contain quite a bit less starch than corn. The best starch sources after corn include corn silage at 30 per cent, wheat midds at 26 per cent, wheat bran at 23 per cent, rice bran at 19 per cent, and corn gluten feed at 16 per cent (DM basis). All other feedstuffs are less than 10 per cent starch. Replacing some of the corn with these feeds will reduce dietary starch to less than 25 per cent which is a common, US ration target. How far can dietary starch be reduced without significantly affecting milk production?

Corn Feed

Corn gluten feed (CGF) is a byproduct of the manufacture of corn sweeteners, corn starch, corn syrup, and corn oil using the wet milling process. The corn starch is used to make ethanol. The CGF consists of the corn bran and a steep liquor (fermented nutrients extracted from water used to soak corn grain) mixed in approximately a 2:1 ratio. CGF is sold in both wet and dry forms. It contains 16 per cent starch, 36.1 per cent NDF, and 23.5 per cent protein, the protein being highly degradable in the rumen. Pricing is substantially lower than corn. CGF may serve as a concentrate, replacing only the corn and soybean meal, or as both a concentrate and a fibre source, replacing both concentrate and traditional forages. CGF generally accounted for 10 to 45 per cent of the diet, although some studies went higher. When CGF was fed at about 20 per cent of ration DM, milk production was decreased statistically in 1 study, increased in 1 study, and remained unchanged in 9 other studies. Cows in these studies were milking between 50 and 90 pounds per day.

Increasing the CGF to 30, 40 or even 57 per cent of the diet did not reduce milk in any study with the exception of Staples et al. In two studies, in which CGF was fed between 38 and 40 per cent of the diet replacing both forage and concentrate starting at calving, average milk yield was increased significantly by feeding the corn gluten feed. As more CGF replaced more corn, the dietary starch concentrations dropped, going as low as 15 per cent in some cases. In spite of lowered starch intake, milk yield did not drop significantly.

The bulk of these studies indicate that dietary starch could be reduced by replacing some corn and protein meals with the digestible fibre found in CGF. Dropping dietary starch from 26 to 21 per cent may be an acceptable compromise. Sample diets were formulated to include CGF at 0 per cent, 10 per cent, and 20 per cent of the diet; starch was reduced concurrently from 25.9 to 23.5 per cent to 20.9 per cent in these rations. As CGF increases in diet, the proportion of extruded soy meal increased to keep the ruminally undegradable protein constant. Degradable protein sources had to be reduced to keep the dietary crude protein from exceeding 17 per cent. Whether this approach will maintain milk production is not known. Milk and feed intake will be maintained when CGF is fed at 20 per cent of dietary DM. But feeding CGF at 10 per cent of the diet may be a preferred approach initially to "play it safe." Some signs that may indicate that the dietary starch has gotten too low include; dropped milk production, stiffer manure, increases in milk urea nitrogen, and loss of body condition.

Another concern is that drying CGF can reduce the digestibility of the protein if the drying temperature is too high. This is determined by analyzing CGF for acid detergent insoluble protein (ADIN). Bernard et al.reported that dried CGF had higher concentrations of ADIN than wet CGF. In recent years, high ADIN may not be as big of a problem in CGF based upon analyses reported by Dairy One. The ADIN values were less than 5 per cent of the nitrogen in the majority of samples; ADIN is usually not considered a problem until it reaches 10 per cent of the total nitrogen in the feed. Commodities need to have some consistent level of nutrient density from delivery to delivery in order to keep the daily ration nutrients consistent for the cows. Dairy One lists on their web site the normal range and standard deviation of the nutrients of most of the feeds they have analysed over the past 6 years.

Corn has an average starch value of 70.6 per cent. The standard deviation is 5.1 per cent. This means that two-thirds of the samples that were analysed ranged between 65.5 per cent and 75.7 per cent which is 5.1 per cent units added to or subtracted from the mean of 70.6 per cent. This also means that one-third of the samples analysed were outside this range. This is a fairly narrow range compared to byproduct feeds.

Starch in CGF samples has a typical range of 8.6 per cent to 24.0 per cent with a mean of 16.3 per cent. This large variation is likely due to the fact that CGF production is not standardized across mills. Likewise, starch in hominy typically ranges between 42.9 per cent and 63.9 per cent with a mean of 53.4 per cent. By feeding these byproducts in smaller amounts, the potential negative effect of nutrient variation on cow performance is reduced. When contracting for these commodities, an acceptable range in variation between loads should be agreed upon ahead of time. Loads delivered outside this agreed-upon-range would then be refused.

Wheat Midds

Wheat midds contain 26 per cent starch and 18 per cent protein. Price can be substantially lower than corn at the right time of the year. Wheat midds have not been evaluated as a substitute for corn in many experiments. In a recent study, wheat midds were fed at about 7.5 per cent of the diet, replacing a combination of corn and soybean meal.

Corn was reduced in the diet from 34.1 to 28.9 per cent. Cows fed wheat midds tended to eat less feed dry matter but milk production and milk composition were not affected. This reduced feed intake may have been because wheat midds have a high water-holding capacity, thus increasing gut fill with fibre. Cows fed wheat midds appeared to have looser manure than those fed more corn and soybean meal.

In a second study, wheat midds were fed at 0 or 22.4 per cent of the diet, replacing 35 per cent of the ground corn and 30 per cent of the soybean meal. Intake of feed dry matter, production of milk, and milk composition were not different between the two groups of cows which averaged 150 days in milk at the start of the study. However the digestibility of dry matter and neutral detergent fibre were lower for cows fed wheat midds. Therefore wheat midds do not appear to be an effective feed replacement for ground corn except for lower producing cows.

Dried Distillers Grains Plus Solubles (DDGS)

Dry corn put through the dry-milling process will produce ethanol, carbon dioxide, and DDGS. Each 56-pound bushel of corn processed through the dry milling process results in 2.85 gallons of ethanol, 18 pounds of carbon dioxide, and 18 pounds of DDGS. Plants using the dry milling process are less expensive to build than the plants using the wet milling process, thus about 75 per cent of the ethanol produced from corn comes from dry milling.

The production of DDGS in the US increased tenfold between the years 1980 and 2000, increasing from 320,000 to 3.5 million metric tons (1 metric ton = 2205 pounds). Production doubled again between 2000 and 2004 to over 7.3 million metric tons. Every dairy cow in the U.S. would need to consume 7.3 pounds per day of DDGS over a 305-day lactation in order to use up this supply, but of course DDGS is also fed to other livestock and is exported.

During the production of ethanol from corn, a syrup product and a cake product are produced. These products are blended together in different proportions to form DDGS. Depending on the plant, the DDGS may be a mixture of syrup to cake ranging in proportion from 35:65 to 55:45. Because of the variation in the ratio of syrup to cake used by different plants to form DDGS, nutrient concentrations will vary from one plant to another.

However the variation looks to be less than that of CGF. Although DDGS is very low in starch, it is an excellent source of protein (30.3 per cent) and fat (13 per cent) and therefore has an energy density very similar to corn. In addition, the ethanol-making process increases the digestibility of the fibre. Unlike CGF, DDGS is a very good source of ruminally undegradable protein (RUP) but if the temperature of the drying process is too high, the protein may be indigestible.

Purchasers of DDGS should analyse for ADIN regularly. Kalscheur et al. reviewed 24 experiments in which 98 comparisons were made between cows fed diets containing DDGS and those not fed DDGS. Cows fed DDGS at up to 30 per cent of the ration produced as much milk as those not fed any DDGS (about 73 lb/day). In spite of decreasing starch content, milk production was maintained. In a study conducted at the University of Florida, whiskey DDGS were fed at 0 per cent or 20 per cent of a 55 per cent concentrate:45 per cent forage diet. The forage was all corn silage in one set of diets and 50 per cent corn silage:50 per cent rhizome peanut silage in another set of diets. The DDGS replaced around 40 per cent of the corn and soybean meal. Milk production was increased from 59.0 to 60.7 pounds per day without changing milk fat test.

The effect of DDGS was the same regardless of whether the forage was totally corn silage or an equal mixture of corn silage and rhizome peanut silage. However, a maximum feeding level of DDGS might be 15 per cent in order to prevent the overfeeding of protein, RUP, unsaturated fat, and phosphorus. Increased corn oil in the DDGS can also lead to decreasing milk fat, so fat content of DDGS should be monitored regularly.

Feeding DDGS at 20 per cent of the diet will increase dietary fat by 2.6 per cent. This will likely be a concern if the diet contains other supplemental fats. Dietary fat should be kept below 6 per cent. The feeding of one pound of DDGS can replace about 0.6 pounds of corn and 0.4 pounds of soybean meal.

Soybean Hulls (SBH)

During the processing of soybeans for oil and meal, the hulls are separated and ground. The resulting soybean hulls consist largely of the outer covering of the soybean so they are high in fibre and contain moderate amounts of protein but very little starch. Protein concentration will depend upon how well the hulls are cleaned. Positive characteristics include a high content of lysine, a low content of phosphorus, and a highly digestible fibre. Fat content can vary widely. The availability of SBH is likely to increase in coming years as more acres are planted to soybeans in order to use soybean oil for biodiesel production. In 13 separate experiments published between 1976 and 2002, SBH (fed at up to 20 per cent of the dietary DM) successfully replaced ground corn or high moisture corn in rations for lactating cows. In most of these studies, the dietary starch levels were probably greater than what is recommended today, based upon the proportion of corn in the diets fed in these studies. Therefore the replacement of corn with SBH still left enough starch to support adequate intake and milk production.

One study fed a more conservative amount of corn in the control diet and deserves further attention here. The control diet was 23 per cent high moisture corn, 26 per cent corn silage, and 26 per cent alfalfa silage. Adding SBH to the diet at 14 per cent of the dietary DM reduced the corn to 9 per cent of the diet.

This dropped the starch from roughly about 25 per cent to 15 per cent, calculated from average values. Diets were fed starting at 8 days fresh.

Lactating cows ate more DM when offered the ration containing SBH but milk production was not changed. Lactating heifers fed SBH ate the same amount of feed and produced the same amount of milk as those not fed SBH. Concentrations of milk fat and protein were unchanged. Replacing 60 per cent of the corn with SBH, so that SBH made up 14 per cent of the ration, was effective to support high milk production in early lactation. Even though dietary NDF increases when SBH replace corn, feed intake has not been decreased. This is in contrast with the well-documented fact that as traditional forage replaces concentrate in the diet, concentration of dietary NDF increases and feed intake decreases. However the NDF in non-forage fibre byproducts like SBH does not have the same physical characteristics as the NDF in traditional, properly chopped forage. The fibre in SBH is highly digestible, very short and has a specific gravity that allows it to move out of the rumen quickly so that SBH fibre does not have the same rumen fill as forage fibre at similar NDF concentrations. In order to use SBH most efficiently, the diet should contain sufficient "effective" fibre from traditional forages.

Corn Silage

Corn silage hybrids differ in the proportion of grain in the total crop. Selecting corn silage hybrids that contain more starch will reduce the amount of ground corn needed. In a test of 55 corn silage hybrids grown in Gainesville in 2006, starch concentration ranged from 22 per cent to 35 per cent of silage dry matter, with an average of 29.3 per cent. Selecting a corn silage hybrid with high levels of starch and digestible fibre, while maintaining high, could be a good strategy to reduce corn costs in the ration. How much could be saved? If a corn silage hybrid had 33 per cent starch instead of 30 per cent, and corn silage was fed at 33 per cent of the ration dry matter, dietary ground corn could be reduced from 18 per cent to 16.5 per cent, which is about 1 lb/day less ground corn. The cost savings would depend upon the relative price of the feedstuff(s) used.

Corn silage contains much more starch than other forages fed in Florida. Feeding more corn silage in the ration and less sorghum silage, cottonseed hulls, or bermudagrass will increase the starch in the diet and allow for less ground corn to be fed. If corn silage inventory allows, increasing the corn silage from 33 to 38 per cent of the diet DM by replacing another forage like sorghum silage, will allow ground corn to be reduced by 1.5 percentage units.

Glycerol

Glycerol is a byproduct made as plant and animal fats are processed to make diesel fuels. Diesel fuel produced from fat burns cleaner, as there is no soot or particulate matter produced. Starting with animal fat to make diesel fuel is more profitable than starting with vegetable oil because animal fats are cheaper. However, diesel made from animal fat may not work as well in colder climates because it "clouds up." This process is expanding worldwide and therefore the supply of glycerol in the future is likely to increase. Ten pounds of glycerol result from every 100 pounds of biodiesel produced.

The purity of this byproduct can vary widely. The more impure the product, the more water, methanol, phosphorus, and potassium it contains. A product with these contaminants is labeled glycerin. The glycerin fed in a German study contained 2.2 per cent potassium and up to 2.4 per cent phosphorus. Methanol is present due to its use in the manufacture of biodiesel. Although ruminal bacteria can detoxify methanol, it can be problematic if present in large quantities. According to the FDA, glycerin is considered a substance that is GRAS (Generally Recognized as Safe) for general purpose use in animal feed, unless methanol is present at concentrations exceeding 150 ppm. The energy content of glycerol is similar to corn when fed in high starch diets. Dietary glycerol would be converted mainly to propionate and butyrate by ruminal bacteria. As a feed ingredient, it could substitute for corn or molasses.

Little research has been done regarding the feeding of glycerol to dairy cows. Lactating dairy cows fed glycerol (1.3 per cent methanol) at either 0 per cent, 2.7 per cent, or 5.3 per cent of dietary dry matter ate the same amount of feed and produced the same amount of milk. Milk composition was unchanged with the exception that milk urea nitrogen concentrations were lower in cows fed glycerol.

Efficiency of fat-corrected milk production was improved from 1.46 to 1.59 and 1.60 lb of FCM yield per lb of feed dry matter intake. Interestingly, animals of other species consumed more water when they were fed glycerol. This may have a benefit to dairy cows managed under heat stress conditions. German researchers fed glycerol up to 10 per cent of ration DM successfully. Glycerol appears to be a good pelleting agent when added at 5 per cent.

New Technologies to improve Ethanol Production

As the ethanol manufacturing industry works at improving the efficiency of conversion of corn to ethanol, different byproducts will become available. Applegate et al. lists the following potential corn byproducts.

- Called the "quick germ quick fibre method," an enzyme is added to the water used to soak the ground corn, causing the germ and fibre to float before the fermentation process begins. The product from this process contains 28 per cent protein, 5 per cent fat, and 25 per cent NDF.
- Collecting the pericarp fibre and germ prior to fermentation by modifying the dry grinding process (drum degerminator) results in a

product that is 24 per cent protein, 8-9 per cent fat, and 28 per cent NDF.

- Removing the fibre by sieving and air aspiration results in a product that is 40+ per cent protein, 15 per cent fat, and 20 per cent NDF.
- Modifying the yeasts used to ferment the sugar to ethanol could allow for an increased concentration of lysine in DDGS, thus giving DDGS a more favourable amino acid profile.

As each mill adopts what they consider the best ethanol-producing technology, the feeding industry will be faced with a variety of byproducts on the market which will need to be properly identified prior to purchase.

With the increased price for corn due to demand for ethanol, farmers will be planting more of their acres to corn. This will reduce the number of acres committed to other crops, such as cottonseed. This shift will likely have a large impact on the market price of a number of feed commodities. The availability and price of each commodity will need to be evaluated for optimal pricing. Some acceptable, alternative feeds may allow some reduction of corn grain in the diet of lactating dairy cows.

FEEDING AND NUTRITION OF PIG

Feeding guinea pig is a fairly simple manner, though there are a few special needs that we should be aware of. Guinea pigs are strict vegetarians, and in addition to their pellets, they will eat fruits and vegetables, bread, and will even nibble on vegetarian dog biscuits. Before getting in to the details, it is important to point out that guinea pigs like a routine, predictable life. Sudden changes in their diet can cause stress, so be careful. If the brands of pellets is changed, gradually by adding a few of the new ones to their old pellets, and slowly switching the mixture over. Feeding excessive amounts of fruits and vegetables at one time can cause diarrhea, particularly if new foods are introduced.

All guinea pigs need water to survive. Though they can obtain a large portion of their water from fresh fruits, vegetables and greens, they will still require water from a water bottle to stay healthy. This water should be changed daily, and the water bottles themselves washed periodically to keep them clean. To prevent algae growth, make sure that the sun does not directly shine on the water bottle for long periods of time. If there is arrangement of tap water bottled water or a water filter system should be considered. Distilled water is not recommended for regular drinking, as it can disturb the osmotic pressure in the body cells, which can potentially lead to osmotic shock.

Be aware that some guinea pigs like to "play" with their water bottle, leaking out large amounts of water by holding the steel ball up in the tube. Others like to try and blow water back up into the bottle, which can contaminate the entire supply. Hay is the mainstay of every guinea pig's diet. Though guinea pig pellets do contain hay, it is not in sufficient quantity to keep guinea pig healthy. Guinea pig must have a fresh supply of hay every day in order to keep its digestive system regular, and as a general rule.

Timothy hay is the best type of hay to feed, though any grass hay will do. If grass hay is not available, a legume hay can be fed, such as alfalfa, though in general this should be avoided as much as possible. Legume hays are high in calcium, and this can lead to bladder stones (uroliths) in some animals. Most pet stores carry dried timothy hay in large and small bags, but it is far better to purchase hay fresh from a feed and garden store, or directly from a farmer. Buying hay by the bale is especially economical, as a bale of hay will literally last for months, and will stay fresh as long as it is stored properly. Some tips on hay storage can be found here.

Guinea Pig pellets

Guinea pig pellets will provide guinea pig with the proper balance of vitamins (save for Vitamin C), minerals and other nutrients. Although it's not necessary to feed pellets to guinea pigs, it is certainly recommended that they be primary feed, after hay. Guinea pig pellets are, in particular, a prime source of protein; obtaining the necessary amounts of this and other nutrients will require careful dietary planning if pellets are not chosen. As a general rule of thumb, adult guinea pigs will eat between one and two ounces of pellets each day. More active animals will eat less, and animals without much stimulation will eat more out of boredom. Nursing and pregnant sows may demand a little more than this, and animals that are given plenty of hay and some fresh vegetables each day might eat a little less. It will be easy to tell if they are fed too much or too little: if there are leftover pellets. If the bowl is empty and they seem to be foraging for food, then they aren't fed enough.

Try to feed at the same hour each day, as guinea pigs like a routine and predictable life. Remove any leftover pellets after an hour or so. Guinea pig pellets are a prime source of protein: a diet that is heavy on pellets will lead to a fat guinea pig. Hay is the only food that they should have access to 24 hours a day. When choosing pellets, be sure to *avoid* the guinea pig "mixes" that contain nuts, seeds and dried fruits. These mixes are high in fats and oils, which can lead to excessive weight gain. Additionally, many of these mixes contain sunflower seeds in their shells. Guinea pigs should *never* be fed nuts or seeds that are still in their shells (peanuts, sunflower seeds, etc.): dozens of guinea pigs in the Pacific Northwest alone die each year from choking on these shell fragments.

One thing that should be noted before we move on: guinea pigs should be fed guinea pig pellets, not rabbit pellets or pellets for another species of animal. Guinea pig pellets are nutritionally balanced for guinea pigs, and feeding them pellets for other animals can result in serious health problems. Though some people may tell that their guinea pigs survive well on rabbit pellets, for instance, it is not trusted that all rabbit food is safe for guinea pigs. Their particular brand may work for guinea pigs, but another brand may lack necessary vitamins or minerals, or have them in insufficient or excessive quantities.

There is no regulation in the pet food industry, so pellets and feeds vary considerably between manufacturers. Don't play Russian roulette with cavies: only feed them pellets that are explicitly and *exclusively* for guinea pigs. It is *extremely important* that guinea pigs receive enough vitamin C each day to prevent scurvy. Guinea pigs cannot manufacture or store vitamin C, so they must obtain it from their diet on a daily basis. Adult guinea pigs will require 10mg of vitamin C each day, and nursing and pregnant sows will require twice that amount (20mg). The caveat: vitamin C, also known as ascorbic acid, breaks down very quickly, so it must be supplied fresh. Although many pellet manufacturers will claim that their feed contains the required amount of vitamin C.

Vitamin C in pellets breaks down during storage, and after about 90 days, there won't be enough vitamin C in the feed to keep guinea pigs healthy. Also note that the food starts breaking down on the shelf, not when the bag of pellets are opened. Unless the manufacturer dates the bag, showing when the feed was packaged, we really have no idea how long pellets have been on the shelf at the local pet store. Since pellets are generally an inadequate source of vitamin C, most owners will either feed sufficient quantities of fruits and vegetables rich in vitamin C each day, or provide vitamin C in the form of supplements added to the guinea pigs' food or drinking water.

Vitamin supplements come in many forms: pre-packaged vitamin supplements for guinea pigs can be bought from pet store, or self grind up vitamin C tablets. In either case, most people prefer to add the vitamins to the animal's drinking water, as that is the only way to guarantee that it will be ingested daily. If the supplement routes are chosen, here are a few pointers:

- Add the appropriate amount of supplement to the amount of water that the guinea pig will drink in a day. Most water bottles work best if they are filled halfway or more, so end up wasting a lot of vitamins. Fortunately, they aren't very expensive. If guinea pig drinks 4 fluid ounces of water a day, and fill water bottle with 8 ounces, then need to add 20mg of vitamin C supplement. This will lead to 4 ounces of wasted water and 10 mg of wasted vitamin C.
- Vitamin C (ascorbic acid) breaks down very quickly in water, so need to add it on a daily basis.
- The chlorine in tap water can actually inactivate ascorbic acid. If vitamin supplements in water is opted, then use tap water that has been standing for at least 24 hours, or filtered tap water from a filter system that reduces chlorine content.

Being vegetarians, guinea pigs will eat many kinds of fruits, vegetables and fresh greens. And, many fruits and vegetables are sources of vitamin C, providing a natural way to meet guinea pig's daily requirement. Calcium and Phosphorus totals for the given food amounts are presented as well. Although these are necessary minerals, and a part of every guinea pig's diet, they are getting them from their pellets, as well. The nutrients guinea pig is getting along with their vitamin C. Too much calcium can cause bladder sludge, a precursor to bladder stones. The proper balance of phosphorus, calcium and vitamin C prevents guinea pigs from needing vitamin D supplements.

Although most guinea pig pellets contain vitamin D as a precaution, there's no guarantee that particular brand of feed does. In addition to providing vitamin C and variance in cavy's diet, fresh greens and fruits contain a significant amount of water. Feeding "wet foods" will aid nursing sows in lactating, and will reduce the amount of water that the average guinea pig will drink from the water bottle. It is especially useful to take lettuce, green peppers and other watery greens on trips to provide moisture for guinea pigs while in the car, since water bottles will leak heavily as the bumps in the road rattle the ball in the drinking tube.

To reduce or completely eliminate pellets from guinea pig's diet, they can survive on water, hay and fresh fruits and vegetables. However, this will require very careful dietary planning on part, to make sure that they receive the necessary nutrients in the proper amounts. Also, be aware that many fresh greens are laxative in action, which means that run the risk of giving guinea pigs loose bowels, or even diarrhea, if the are fed too much at one time. If runny droppings are noticed immediately cut fresh greens out of their diet and feed dry foods until the feces returns to normal. Vegetables that are not laxative (such as carrots) may still be fed. Guinea pigs will not eat what they don't like: some cavies have very discriminating tastes, while others will eat anything that put in front of them. Each pig has its own preferences, and it may take some time to figure out what he or she likes best. Given below is a list of foods that guinea pigs can/will eat. This list is neither comprehensive nor complete.

Apples, bananas, bread (slightly stale and crunchy, but not moldy), broccoli, carrot greens, carrots and baby carrots, celery (cut into small pieces first), cilantro, cucumber, dandelion greens, grass, green and red bell peppers, green leaf and romaine lettuce, kale, kiwi, mustard greens, oats, oranges, parsley, raspberries, spinach, tomatoes. When feeding "wild" greens, such as grass and dandelion greens, make sure they have not been sprayed with chemicals, or contaminated by droppings or urine from other animals, such as cats, dogs and birds. The dangers of pesticides are obvious, and feces can carry any number of parasites which can be transmitted to cavies eating contaminated greens. Even if dogs and cats are dewormed regularly, their feces can contain protozoa/bacteria which can cause debilitating diarrhea in a guinea pig.

Some foods to avoid are listed below:

- Long celery stalks
- Iceberg lettuce
- Any shelled nuts or seeds
- Raw beans
- Rhubarb

There is some confusion as to whether or not potato peelings are good or bad: some books indicate that they are poisonous to guinea pigs, while others say that they are okay in small amounts. The truth is that potatoes are *okay*, however, any *green* in a potato *is poisoinous*. If potato peelings are given, make sure there are *no green spots* anywhere in the portions. Also, some people may recommend yogurt in small amounts, for it's bacteria-growing properties (to aid the digestive system). Instead, look into acidophilus powders or liquids: acidophilus is also a bacteria growing culture, and it';s likely to give better results since less of it needs to be fed.

There are several commercially available treats on the market that are aimed at guinea pigs. Berry flavoured "crunchies" and flavoured chew sticks are popular among some owners and guinea pigs, and are generally safe. Some owners also have experimented with vegetarian dog biscuits and dry cereals such as Cheerio's. However, the guinea pig "treat sticks", which are also available commercially. These sticks are essentially seeds and nuts that are held together with honey. Most of the treat sticks on the market contain sunflower seeds that are still in their shells, which is a big no-no for guinea pigs due to the choking hazards.

FISH FEEDING

Although fresh animal by-products such as liver and blood are available in limited quantities and could be used by a small-scale trout farmer (< 5 tonnes fish/year) within an in-house produced moist diet, a dry diet formulation and feeding regime is recommended for the Project fish farms.

Three dietary formulations are recommended for use within three distinct feed lines, namely starter, fingerling and production diets. Table shows the recommended nutrient levels and dietary formulations for the three feed lines, respectively. The formulations are conservative in that emphasis has been placed on using high dietary inclusion levels of known quality ingredients (ie. such as fishmeal; Table) rather than experimenting with unknown cheaper alternatives such as rapeseed meal. For example, although oilseeds (ie. soybean meal, rapeseed meal) and animal by-product meals (ie. meat and bone meal, blood meal) can and have been successfully used at high dietary inclusion levels within practical salmonid rations (Table), their nutritional success is dependent upon the individual manufacturing process used to produce them. Thus, since the level of technology available in Qinghai Province to manufacture these ingredient sources is still in its infancy, and the proximate composition of these ingredient sources is highly variable, their inclusion level within the proposed diet lines has been kept to a minimum.

Providing fish with the right type of food in the right proportions is a very important task. This can be much more of a challenge in a saltwater tank than in a freshwater tank, because most marine fish are still wild-caught, and their natural diet can be difficult to maintain in captivity. With the right information, however, most marine fish will learn to eagerly anticipate their feeding time and will voraciously devour their food.

Fig. Fish Food

A good rule of thumb to follow is that fish should be fed no more than they can consume within a five minute period. Any food that is left behind after feeding time should be removed before it has a chance to begin decomposing. If it is found that fish are consuming the food in far shorter a time than five minutes and seem anxious for more, the amount that are fed must be increased.

On the other hand, if fish are healthy, but they are only consuming about half of what they are fed, probably need to start feeding less. Always keep in mind that much more harm can be done to fish by overfeeding than by underfeeding. In the wild, reef fishes will browse for food throughout the day. Their natural habitat as closely as possible, it is recommended that fish must be fed several small meals every day. At the very least, feed two medium-sized meals per day, one in the morning and one at night.

Types of Fish Food

The kind of food to feed depends on the kind of fish are feeding: herbivore, carnivore, or omnivore.

Herbivores

It is common for fish that graze on algae in the wild to suffer in captivity because they don't receive the appropriate type of food to keep them healthy. Many of these fish are fed terrestrial greens like spinach and romaine lettuce, but these are not appropriate foods for these fish that are solely algae eaters.

They should be fed either dried or fresh algae. Dried algae is sold in paper-thin sheets that can be broken down into appropriate portion sizes, depending on the number of fish are feeding. Fresh algae can be obtained by placing a glass container, with one rock in aquarium, on a sunny windowsill. Wait for the rock to grow a nice layer of algae, and then place it back in the aquarium for fish to graze on. Herbivorous fish can also be fed one of the frozen fish foods that is formulated especially for herbivores. This food should contain a variety of marine algae, vegetable matter, and seafood. Whether herbivorous fish are fed dried algae, fresh algae, or a frozen food formula, herbivorous eaters are grazers and should have food available to them at all times, unlike most other fish.

Carnivores

Meat-eating marine fish can enjoy a variety of fresh, frozen, freeze-dried, and live foods. Fresh foods suitable for fish include many seafood items, such as shrimp, scallops, mussels, clams, and some fish. If seafood is fed to fish, cut it into bite-sized pieces (or serve it whole to larger fish). Fresh seafood can be frozen and then thawed as needed.

Frozen foods are available from most fish shops and other sources. These offerings may include brine shrimp, bloodworms, and daphnia. They may be in the form of a slab that can be broken into smaller pieces or small, ready-to-go cubes.

To feed frozen foods to carnivorous fish, thaw a small portion in aquarium water, and then drop it in the tank. Several freeze-dried foods for carnivorous fish are readily available, such as plankton, krill, brine shrimp, and bloodworms. Freeze-dried foods are foods that have been completely dehydrated, yet they maintain most of their nutritional value. They have a relatively long shelf life, particularly when compared to fresh and frozen foods. Freeze-dried foods should be used as supplements for carnivorous fish. Live foods are almost always a favourite of fish. Adult brine shrimp, bloodworms, and Mysis shrimp are among the better choices. Earthworms are also a great live food choice because they come in all sizes, fish love them. Be sure to thoroughly rinse them before feeding them to fish.

Omnivores

The majority of marine fish are omnivorous, which means they need to eat both meat- and plant-based foods. One easy option for omnivorous eaters is commercial fish food, such as flakes or pellets. However, offering a varied diet will give healthier, more colourful fish. Try to feed fish two feedings a day of commercial fish food and one feeding of a meaty item, such as bloodworms or chopped fish. The food offered must be alternated in addition to commercial food to keep the fish from getting bored and to make sure all of their nutritional needs are met.

ANALYSIS OF FEED

In order to formulate, produce and market animal diets it is necessary to gain information about substance classes (i. e. that contribute to the energy content) and special compounds (i. e. that influence digestibility). Different chemical, physical and biological methods have been developed to obtain the information needed.

Sampling

The most important precondition when analyzing feedstuffs and diets is the sampling procedure. Thereby the person taking the samples has to keep the following principles in mind:

1. The sample taken must be representative for the entire lot.
2. Samples are to be taken randomly from several different points of the lot. Subsequently the samples are then mixed to a single blend to produce a collective sample, which again is divided into several representative laboratory samples for analysis.
3. Sampling equipment should not be able to influence the sample taken, i. e. via contamination or sedimentation.
4. Feed samples need to be stored in a manner that ingredients will not be altered (temperature, oxygen, sun light etc.) before analysis.
5. A sampling report should be prepared in order to assign the sample correctly after analysis.

Proximate Analysis (Crude Nutrient Analysis)

The proximate or Weende analysis of feed is a quantitative method to determine different macronutrients in feed. Basically it is the partition of feed compounds into six categories by means of common chemical properties. The categories are moisture (crude water), crude ash (CA), crude protein (CP), ether extracts (fats or lipids; EE), crude fibre (CF) and nitrogenfree extractives (NFE).

The feed sample is initially dried at 103 °C for 4 hours. The weight loss of the sample is determined and the crude water fraction is calculated. Ashing the sample at 550 °C for 4 hours removes the carbon from the sample, *viz.* all organic compounds are removed. Again calculating the weight loss of the feed sample from the dry matter to crude ash (CA) content mathematically determines the organic matter fraction.

The nitrogen content of the food is the basis for calculating the crude protein (CP) content of the feed. The method established by Kjeldahl converts the nitrogen present in the sample to ammonia which is determined by titration. Assuming that the average nitrogen content of proteins is 16 per cent multiplying the nitrogen content in per cent obtained via Kjeldahl analysis with 6.25 gives an approximate protein content of the sample. Alternatively the Dumas method can be applied to measure CP. Fats and lipids are extracted continuously with ether, after evaporation of the solvent the residue remaining is the ether extract (EE) fraction. The carbohydrates in a feed sample are retrieved in two fractions (CF, NFE) of the proximate analysis. The fraction, which is not soluble in a defined concentration of alkalis and acids, is defined as crude fibre (CF). This fraction contains cellulose, hemicellulose and lignin. Sugars, starch, pectins and hemicellulose etc. are defined as nitrogen-fee extractives (NFE). This fraction again is not determined chemically it is rather calculated by substracting CP, EE and CF from organic matter.

In recent years the over 100 year old proximate system has been advanced and improved. Especially the imprecision of CA, CF and NFE as well as CP had been criticized. Modern methods to determine the exact composition of the CA fraction via atomic absorption spectroscopy and the CP fraction via amino acid analyzers, near infrared spectroscopy (NIRS) etc. have been established. Improving the information gained from analysis of feedstuffs and diets also involves the determination of sugars and starch (polarimetric methods) contained in the NFE fraction of the proximate analysis. Van Soest developed a procedure to detect the different components of the cell wall. This helps specify the CF and NFE fraction.

Thereby the complete amount of cell wall components is obtained by digesting (boiling) the feed sample in a so called neutral detergent solution and results in the neutral detergent fibre fraction (NDF). The residue after digestion in a solution with sulfuric acid is called the acid detergent fibre (ADF) and contains mainly cellulose and lignin. Finally the remaining sample is treated with a sulfuric acid with an even higher concentration resulting in a decomposition of cellulose leaving mainly lignin. This fraction is called acid detergent lignin (ADL). The combination of the proximate analysis with these modern methods allows for a detailed feedstuff analysis.

UNDERSTANDING FEED ANALYSIS

Feed costs represent the largest annual operating cost for most commercial cow-calf enterprises. In order to maintain an optimum balance between feed costs and production, feeds must be analyzed and these analyses used to formulate rations and (or) supplements. Feedstuffs vary widely in nutrient concentration due to location, harvest date (maturity), year, and other management practices. Tabular values may be used if necessary, but it is important to remember that they are average values and that significant variation exists. On a dry matter basis, energy can easily vary ± 10 per cent, crude protein ± 15 per cent, and minerals by a much greater margin.

Once a feed sample has been collected properly, it can be analyzed for nutrients. Most commercial laboratories offer standard feed tests for forages, grains, or total mixed rations. Analyzing cattle feeds for moisture, protein, and energy is recommended. Furthermore, you may wish to identify key minerals or minor nutrients of interest. Typically, results are reported on an as-is and dry matter basis. Nutrients should always be balanced on a dry-matter basis because nutrient requirements for beef cattle are reported on a dry-matter basis. After formulation on a dry-matter basis, values can be converted to an as-is basis (using the moisture content of the feed) to determine the actual amount of feed (as-is) that should be fed.

Feedstuffs can be analyzed using traditional wet chemistry technique or near infrared reflectance spectroscopy (NIR). Samples can be analyzed more quickly, and usually cheaper, using NIR. However, NIR is only useful for feedstuffs and ingredients that have been well characterized using wet chemistry. Therefore, be sure to ask the laboratory if their database for your particular sample is extensive enough to ensure accurate results, particularly if you are analyzing less common feedstuffs.

MOISTURE

- *Dry Matter* (DM): Dry matter is the moisture-free content of the sample. Because moisture dilutes the concentration of nutrients but does not have a major influence on intake (aside from severe deprivation), it is important to always balance and evaluate rations on a dry-matter basis. Digestible Dry
- *Matter* (DDM): Calculated from acid detergent fibre (ADF; see below); the proportion of a forage that is digestible.

PROTEIN

- *Crude Protein* (CP): Crude protein measures the nitrogen content of a feedstuff, including both true protein and non-protein nitrogen. In ruminants, evaluation of the fraction that is degradable in the rumen, degradable intake protein (DIP), versus the rumen-undegradable fraction, undegradable intake protein (UIP), is also important. However, the rumen degradability of protein is not measured in most commercial labs. Therefore, it is recommended that rations be formulated using analyzed CP values and average values for DIP and UIP that can be found in the 1996 National Research Council Nutrient Requirements of Beef Cattle.
- *Degradable Intake Protein* (DIP): The fraction of the crude protein which is degradable in the rumen and provide nitrogen for rumen microorganisms to synthesize bacterial crude protein (BCP) which is protein supplied to the animal by rumen microbes. DIP also includes non-protein nitrogen found in feeds or ingredients.
- *Undegradable Intake Protein* (UIP): The rumen-undegradable portion of an animals crude protein intake. Commonly called "bypass protein" because it bypasses rumen breakdown and is mainly digested in the

small intestine. Bypass protein is utilized directly by the animal because it is absorbed as small proteins and amino acids.

- *Metabolizable Protein* (MP): MP is protein that is available to the animal including microbial protein (BCP) synthesized by the rumen microorganisms and UIP.
- *Heat Damaged Protein or Insoluble Crude Protein* (ICP): Nitrogen that has become chemically linked to carbohydrates and thus does not contribute to either DIP or UIP supply. This linkage is mainly due to overheating when hay is baled or stacked with greater than 20 per cent moisture, or when silage is harvested at less than 65 per cent moisture. Feedstuffs with high ICP are often discolored and have distinctly sweet odours in many cases. When the ratio of ICP:CP is 0.1 or greater, meaning more than 10 per cent of the CP unavailable, the crude protein value is adjusted. Adjusted crude protein (ACP; see below) values should be used for ration formulation.
- *Adjusted Crude Protein* (ACP): Crude protein corrected for ICP. In most nutrient analysis reports, when ACP is greater than 10 per cent of CP, the adjusted value is reported. This value should be used in formulating rations when ICP:CP is greater than 0.1.
- *Digestible Protein* (DP): Reported by some laboratories, do not use without the guidance of a nutritionist. Digestible protein values are not needed for most ration formulation because nutrient requirements and most formulation tools are already adjusted for protein digestibility. Furthermore, protein digestibility is influenced by external factors.

FIBRE

- *Crude Fibre* (CF): Crude fibre is a traditional measure of fibre content in feeds. Neutral detergent fibre (NDF) and acid detergent fibre (ADF) are more useful measures of feeding value, and should be used to evaluate forages and formulate rations.
- *Neutral Detergent Fibre* (NDF): Structural components of the plant, specifically cell wall. NDF is a predictor of voluntary intake because it provides bulk or fill. In general, low NDF values are desired because NDF increases as forages mature. Acid Detergent Fibre (ADF): The least digestible plant components, including cellulose and lignin. ADF values are inversely related to digestibility, so forages with low ADF concentrations are ususally higher in energy.

ENERGY

- *Total Digestible Nutrients* (TDN): The sum of the digestible fibre, protein, lipid, and carbohydrate components of a feedstuff or diet. TDN is directly related to digestible energy and is often calculated based on ADF. TDN is useful for beef cow rations that are primarily forage. When moderate to high concentrations of concentrate are fed, net energy (NE, see below) should be used to formulate diets and predict animal performance. TDN values tend to underpredict the feeding value of concentrate relative to forage.

- *Net Energy* (NE): Mainly referred to as net energy for maintenance (NEm), net energy for gain (NEg), and net energy for lactation (NEl). The net energy system separates the energy requirements into their fractional components used for tissue maintenance, tissue gain, and lactation. Accurate use of the NE system relies on careful prediction of feed intake. In general, NEg overestimates the energy value of concentrates relative to roughages.
- *Ether Extract* (EE): The crude fat content of a feedstuff. Fat is an energy source with 2.25 times the energy density of carbohydrates.
- *Relative Feed Value* (RFV): A prediction of feeding value that combines estimated intake (NDF) and estimated digestibility (ADF) into a single index. RFV is used to evaluate legume hay. RFV is often used as a benchmark of quality when buying or selling alfalfa hay. RFV is not used for ration formulation.
- *Relative Forage Quality* (RFQ): Like RFV, RFQ combines predicted intake (NDF) and digestibility (ADF). However, RFQ differs from RFV because it is based on estimates of forage intake and digestibility determined by incubating the feedstuff with rumen microorganisms in a simulated digestion. Therefore, it is a more accurate predictor of forage value than RFV. Neither RFV nor RFQ are used in ration formulation.

FEEDING GOATS

Goats are browsers, like deer, which means they prefer trees, bushes, and woody weeds; rather than standing still and eating grass down to the roots, they like to stay on the move, eating a bit of this and a bit of that. Goats can learn to graze a pasture, but don't expect it to be "mowed." The grass helps supplement the goats' diet, but low grazing also can spread parasites. Goats have specific nutritional needs, only some of which are met by the plants on your farm that they browse on. You have to provide feed for the needs that can't be met by browsing. Unless you have a lot of property with a variety of browse, feed will be your biggest expense in raising goats. Don't scrimp on goat feed — it will pay dividends in good health, milk production, and lower veterinary bills.

Feeds for goats can be divided into several different categories: forages, grains, protein supplements, minerals and water. Depending on what levels of each category that is fed. It will be found that goats will gain about.25 to.5 a pound each day. Goats will readily eat a variety of forages. These include pasture, hay, haylage, and brush. Any species of forage is acceptable for goats, although be careful of feeding a lot of alfalfa to wethers and billies because the higher calcium levels can lead to urinary calculi or kidney stones. Goats are very useful in many operations for cleaning up unwanted brushy areas. Most brushy areas will only be able to survive about 2 years with intensive defoliation. Forages are an important part of a ruminant's ration because the high fibre is necessary to keep the digestive system healthy. Grains provide the energy portion of a ration for goats. A variety of grains are available, with corn, oats, wheat and barley being the most widely used. Limit the amount of wheat in the ration to no more than 2 per cent to prevent any digestive problems. When starting goats on a feedlot ration, be sure to gradually introduce them to grain.

Feeding high levels of grain suddenly can cause bloating and overeating disease. Disease can be prevented by vaccinating with type C and D perfringens. This is often available in one vaccine, along with tetanus. Any male goats that are castrated should receive a dose of tetanus vaccine. Castration is a personal consideration as many ethnic groups prefer to purchase intact males.

Protein Supplements

Market goats should be fed a grain ration that runs between 14 per cent and 18 per cent protein. To reach these protein levels, we need to include some sort of protein supplement. Typical supplements will include soybean meal, roasted soybeans, cottonseed, brewers or distiller's grains, or flour mill by-products. Beware of using any dairy protein supplements because of the amount of copper in the supplements. Many nutritionists still disagree over the amount of copper required in a goat ration, but care should still be given to overfeeding this nutrient.

Minerals

One of the best ways to provide minerals to a market goat is through a commercially mixed salt and mineral mix. Be sure to purchase a mix formulated specifically for goats. Allow goat's free access to a mineral feeder that contains the mineral mix at all times. Minerals are especially important for helping to ensure that goats remain healthy. These mineral mixes will often include some of the essential vitamins for goats as well.

Water

Water is often one of the most overlooked aspects of livestock feeding operations. Clean, fresh water should be available at all times for all species of livestock. Animals that have adequate amounts of water are more likely to stay healthy as well as have a faster growth rate. Goats will often be able to supply all of their water needs through dew and lush pasture. However, it is still in their best interest to have a supply of water available to them at all times.

FEEDING YOUR GOAT HERD

Feed is the largest cost associated with raising goats. It can affect herd reproduction, milk production and kid growth. Late gestation and lactation are critical periods for doe nutrition. Nutrition level determines kids growth rates. Goats not receiving adequate nutrition are more prone to disease and will fail to reach their full potential. Goats require energy, protein, vitamins, fibre and water. Energy is the most limiting nutrient, while protein is the most expensive. Imbalances of vitamins and minerals can limit animal performance and lead to various health problems. Fibre maintains a healthy rumen enviroment and prevents digestive disturbances. Inadequate water intake can cause various health problems.

WATER - Water is the cheapest feed ingredient and often the most neglected. Goats should have free access to clean fresh water at all times. It is critical that they have an adequate supply of water in the Winter months. A mature goat will consume between 3/4 to 1 1/2 gallons of water per day.

PASTURE AND BROWSE- Pasture and browse are the most economical source of nutrients for goats. Pasture tends to be high in energy and protein in its vegetative state. As the pasture plants mature the palatability and digestibility decline, making it important to rotate pastures to keep plants in a vegetative state. Some of the best pastures for goats are Bahiagrass, millet, sorghum, sudan grass and a mixture of a grain, grass and clover. During the early part of grazing season, browse tends to be higher in protein than ordinary pasture. Goats are natural browsers and select plants at their most nutritious state. Goats that browse have less problems with internal parasites.

HAY- Hay is the primary source of nutrients for goats during the winter months. Hay varies in quality and the only way to know the nutritional content is to have it analyzed by a forage testing laboratory. Legume hays- alfalfa, clover, lespedeza- tend to be higher in protein; this also depends on which cutting it is.

Briers and such - Goats love eating multifloral rose bushes, green briers, poison ivy and just about every type of unwanted brush located on your property. Many herds are used specifally to "clean" up overgrown property, which can be more economical than hiring people to do the job.

VITAMINS AND MINERALS- Many mierals are required by goats. The most important are salt, calcium and phosphorus. Vitamins are needed in small amounts. Goats require vitamins A, D, and E. Offer a pre-mix of loose minerals free choice; goats will consume more if loose is available.

GRAIN- It is often necessary to feed grains when forage alone cannot provide enough nutrients. CREEP FEEDING and supplemental feeding of kids does increase growth weight, but should only be done to the extent that increases profits. There are two types of feed- carbon and protein. Carbon or 'energy' feeds include cereal grains- corn, barley, wheat, oats, milo, and rye-various by products feeds such as fat, soybean hulls and wheat middlings. Protein supplements may be of animal or plant origin and include soybean meal, cottonseed meal and fish meal. 14 per cent CP to 16 per cent CP will usually fill the nutritional requirement for various classes of goats.

OTHER FEED ADDITIVES

For herds that are prone to urinary calculi, ammonium chloride can be added to prevent the disease. Another common practice is to add a coccidiostat to the ration to prevent as well as control coccidiosis. Common additives used to prevent coccidiosis include Rumensin, Bovatec or Deccox (decoquinate). Lastly, no matter what the ration, be sure to provide enough feeder space for all the goats to eat. Plan for at least 1 foot of feeder space for each goat. And, don't forget to clean feeders and waterers on a regular basis.

Sound feeding practices and good housing facilities result in optimum growth and high milk production and contribute to the good health and comfort of goats.

When feeding goats, keep these objectives must be in mind:

- Feed a young animal enough energy for growth, and feed a mature animal enough energy to maintain a fairly constant body weight;
- Provide enough protein, minerals and vitamins in a balanced feeding programme to maintain a healthy animal; and
- Offer does enough extra food during gestation and lactation for fetus development and milk production.

Optimum growth, good health and high milk production are the results of sound feeding practices. Dairy goats are not unique in their body requirements; they will respond to good nutritional practices. Digestible fibre is especially important in dairy goat diets. Too much grain in relation to forage does not foster good ruminant action and is a costly feeding practice. When feeding, keep minerals and trace mineral salt separate. Always feed them separately.

Feed hay in a rack that will not permit wasting. One recommended type of feeder is a keyhole feeder. Do not overlook forage testing as an economical way to feed the correct ration to herd.

Feeding Kids and Yearlings

Kids are born with no natural protection from disease. The first milk (colostrum) from the mother offers protection and gets the digestive system working. To be most effective, it must be fed before disease-producing organisms enter the mouth and digestive tract. Wash the fresh doe's udders and teats with warm water.

Hand milk half a cup of colostrum and feed it to the kid within 15 minutes of birth. This is the best way to ensure that the newborn receives some milk and to provide it with the most protection from organisms present on the skin of the doe. Complete the first milking and store the colostrum for later feedings if hand feed is elected. Otherwise, permit the kid to nurse at its convenience following the first hand feeding.

Clean the feeding utensils immediately after each use. The same cleaning procedures should be followed for washing milk-handling equipment. Milk replacer may be fed from the fourth day; it should contain at least 20 per cent protein, 20 per cent fat and be free of vegetable products. A lamb or high-quality calf milk replacer is recommended. Provide hay and grain at one to two weeks of age.

Wean from milk when grain intake reaches 1/4-pound daily and kids are readily consuming hay. Make sure the solution is thoroughly mixed. Mix a fresh mixture daily and feed in place of milk. Feed as soon as diarrhea is noticed. Use for 1-1/2 or 2 days, then return to the regular milk diet. After four to six months of age, the kids may be fed a ration similar to that of the milking herd. Good hay and 1/2 pound of grain per day should provide an ample growth rate. Poor hay may require 1 to 1-1/2 pounds of grain daily.

Feeding the Milking Herd

If milk production is important, feed maximum amounts of high quality hay balanced with a grain ration containing enough protein, minerals and vitamins to support production and animal health. Grass or legume hays are equally acceptable. As the percentage of legumes is increased, the need for protein in the grain mix is reduced.

To determine the amount of grain to feed, consider level of milk production, amount and quality of forages consumed, appetite and state of fleshing. Thin, high-producing does should have access to all the hay they can eat plus grain to the limit of their appetite. Does in mid-lactation that are in good flesh should have all the hay they will eat plus 1 pound of grain for each 3 pounds milk produced. Late lactation does may not need more than 1 pound of grain for each 5 pounds of milk.

Feed a grain ration formulated for a milk-producing ruminant (dairy cows). Rolled or cracked grain is more palatable than ground grain. Because of palatability problems, urea is not recommended. Some commercial cow feeds may contain byproduct ingredients unpalatable to goats. Wet molasses is more palatable than dry molasses. Beer or citrus pulp is a valuable source of fibre, especially if the available hay is of low quality.

If the doe is not thin, reduce the amount of grain to 1/2 to 1 pound per day. Feed her all the forage she will eat. Hay fed during the dry period may be of lower quality, but if so, the grain ration should contain additional protein. Browse, leaves and weeds are often useful to recondition the stomach. If the dry ration differs from the milking ration, be sure to change to the milking forage and grain ration two weeks before the doe freshens. Most bucks do not need more than a pound of grain per day plus forages. Don't let them grow fat. Adjust grain upward or downward accordingly. Always feed full forage.

SHEEP DIET

Mostly sheep eat grass, clover, forbs, and other pasture plants. They especially love forbs. It is usually their first choice of food in a pasture. A forb is a broad-leaf plant other than grass. It is a flowering plant. Forbs are often very nutritious. As compared to cattle, sheep eat a greater variety of plants and select a more nutritious diet, but less so than goats.

Sheep are exclusively herbivorous mammals. Like all ruminants, sheep have a complex digestive system composed of four chambers that allows them to break down otherwise inedible cellulose from stems, leaves, and seed hulls in to complex carbohydrates. When sheep graze, vegetation is chewed in to a mass called a bolus, which is then passed in to the first chamber: the rumen. The rumen is a 5-10 gallon (18.9-37.8 litre) organ which ferments feed via a symbiotic relationship with the millions of bacterium, protozoa, and yeasts which are collectively called flora.

The bolus is periodically regurgitated back in to the mouth as cud for additional chewing and salivation, this process usually takes around six hours. Cud chewing is an evolutionary adaptation that allows ruminants to graze more quickly in the morning, and then fully chew and digest feed later in the day. This is beneficial because grazing, which requires lowering the head, leaves sheep more vulnerable to predators, while cud chewing does not. During fermentation, the rumen produces excess gas which must be expelled; disturbances of the organ, such as sudden changes in a sheep's diet, can cause potentially fatal conditions such as bloat. After fermentation in the rumen, feed passes in to the reticulum and the omasum; special feeds such as grains may bypass the rumen altogether.

Following the first three chambers, food moves in to the abomasum for final digestion before processing by the intestines. The abomasum is the only one of the three chambers directly equatable with the usual mammalian stomach, and is accordingly sometimes called the "true stomach".

Sheep follow a diurnal pattern of activity, beginning to feed at dawn and throughout the day until dusk, stopping sporadically to rest and chew their cud. Ideal pasture for sheep is not lawn-like grass, but an array of grasses, legumes and forbs. Types of land where sheep are raised vary widely, from pastures that are seeded and improved intentionally to rough, native lands. Common plants toxic to sheep are present in most of the world, and include (but are not limited to) oak and acorns, tomato, yew, rhubarb, potato, and rhododendron. Sheep also may perish from consuming man-made substances.

Sheep are largely grazing herbivores, unlike browsing animals such as goats and deer that prefer taller foliage. With a much narrower face, they can crop plants very close to the ground and can overgraze a pasture much faster than cattle. For this reason, many shepherds utilize some type of managed grazing, whereby a flock is rotated through multiple pastures, giving plants time to recover.

Paradoxically, sheep can be both the cause of and solution to one of the major ill effects of overgrazing: the spread of invasive plant species. By disturbing the natural make-up of pasture, sheep and other livestock can pave the way for invasive plants. But sheep also prefer to eat invasives such as cheatgrass, leafy spurge, kudzu and spotted knapweed over native species such as sagebrush—thereby making grazing sheep an effective measure in restoring native pastures.

Other than forage, the only other common staple feed for sheep is hay, most often during the winter months. The ability to survive solely on pasture (even without hay) varies with breed, but all sheep are capable of doing so. Also included in most sheep's diets are minerals, either in a trace mix or in licks.

Naturally, a constant source of potable water is also a fundamental requirement for sheep. The amount of water needed by sheep fluctuates with the season and the type and quality of the food they consume. When sheep feed on large amounts of new growth and there is precipitation (including dew, as sheep are dawn feeders), sheep need less water. When sheep are confined or are eating large amounts of cured hay, more water is typically needed. Sheep also require clean water, and may refuse to drink water that is covered in scum or algae.

COOLING PONDS FOR DAIRY CATTLE

Cattle have used water for cooling probably for as long as there have been cattle. The last survey done in Florida reported that 30 per cent of the dairies had cooling ponds. The reader should be aware that cooling ponds, if not maintained properly, could harbour infectious diseases such as Leptospirosis and many mastitis organisms. The water usage is very high to fill and keep full.

Environmental concerns that must be addressed are the run-off from ponds and ground seepage. Every state seems to have many regulations that must be addressed before building cooling ponds. The benefits of cooling ponds and some suggestions on building and maintaining them. It is the dairymen's responsibility to use them properly and to be aware that ponds can harbour diseases and the pond water must be dealt with according to state and local laws.

What is a cooling pond for dairy cattle? For this discussion they are a man made hole in the ground with a constant inflow of clean fresh water. Usually the ponds drain or overflow from the top. If a series of ponds is used, the overflow water may be run into a retention pond to be irrigated on crops or enter the waste management system. Cooling ponds may also be fed from springs or artesian wells. So cooling ponds for dairy cattle should have constant fresh water in and a constant outflow of water to some place for proper disposal.

Cement ponds may also be built. These may drain from the bottom and can be used to flush feed lanes or barns. Bottom draining will help remove some of the solids that have settled to the bottom of the pond. Care must be taken on the entrance and exit to cement ponds.

These may be very slippery if the cement is not cross-grooved. A stagnant pond without fresh water entering, a lagoon, a wallow, and a mud hole do not classify as a cooling pond for dairy cattle.

Site Selection

A site near the feed and drinking water, ideally with some form of shade for cows, is best. The idea is for cows to cool, leave the pond and goes eat and drink, lay down in the shade and then do it all over again. If cows do not have feed and drinking water near, they will stay in the pond and not eat and we have cool wrinkled cows that do not give much milk. Cooling ponds should be located close to a water supply as fresh water needs to enter the ponds constantly. There must be a way to dispose of run-off water properly. This will vary by areas of the country. Some slope from the ponds to the waste management area would be helpful, so pumps need not to be used. Cooling ponds should not be too far from the milking parlor as long walks in the sun increase heat stress. Exit lane sprinklers help make the long walk more comfortable.

Concrete Ponds

Concrete ponds will work and have been used but they are expensive to build. However, they are easier to maintain, as the entrances and exits do not have to be rebuilt. The finish of the concrete on exit and entrance slopes should be cross-grooved to prevent cows from falling. A 1:8 slope would be a safe slope if concrete ponds are constructed.

Sizing Cooling Ponds

There is no official listing for cooling pond size. If a holding area requires 15/ft^2/ cow and cows are driven into a pond and taken out as a group then 15/ ft^2/ cow might be adequate. If 50/ft^2/ cow is recommended for most general animal space, then this might be the correct figure for a cooling pond.

This amount of space would allow cows to enter and exit the pond at all times when all cows were not using the pond at the same time. If 50/ft^2/ cow for a pond is used, then 5000/ft^2 is then needed for the 100 cow group.

Pond Shapes and Depths

Using the sample of 5000/ft^2/100 cows a pond could be 50' wide × 100' long, or a circle 80' in diameter (A= (Pi) r^2; A= 3.14 × 40 = 5024ft^2). If a 50' ×

100' pond is used and the sides are fenced so the cows can only enter or exit at each end of the pond, the amount of dirt falling into the pond will be reduced. If a circular pond is made, cow carpet should be used to eliminate the problem of dirt falling in the pond. As cows can swim, pond depth is not a problem.

Deeper ponds may allow the settling of organic matter on the bottom of the pond. This organic matter may not be disturbed by cows walking through it as they will swim over the top of it. Preference should be given to depths of 3' or 4' as it seems to cool the cows and let them stand and walk in and out at will. While deep ponds allow cows to submerge, they do not seem to like to float too long so they go to where they can stand and may block cow traffic. A circular pond that is deep in the middle lets cows dunk themselves while others enter and exit the pond at will.

Water Usage

A 100 cow group with 50/ft^2/ cow could have a 50' wide × 100' long × 3' deep cooling pond (it may be 4' in the middle and slope in and out). A cooling pond 50' wide × 100' long × an average 3' deep=15000/ft^3. There are 7.48 gallons/ft^3, thus the pond requires 112,200 gallons of water.

The pond must be filled and then kept fresh with running water at all times while in use. A 1" pipe 300 ft. long and water pressure at 35psi will result in a flow rate of about 23.5 gpm. At this rate it would then take 80 hours to fill the pond initially.

The dairy water usage would then be 1,410 gph × 24 hours or 33,840 gallons of water per day. This fresh water must be pumped into the pond and there is going to be close to that amount of run off water that will exit the pond for disposal.

Water Run-off

Disposal of run-off water must be in accordance with whatever local, state, and federal laws that applies. Many dairies run the water into the waste management system, while others add it to retention ponds to be used for irrigation.

Shading over Ponds

Experience has led to discontinued use of shade over ponds as the cows never leave the ponds. Remember, the object of this exercise is to increase dry matter intake and cows will not eat much soaking in the pond.

Pond Maintenance

At least once a year, usually during the winter, the pond should be fenced and the water removed by pumping. The mud and manure on the bottom of the pond should be removed and the pond left empty for the sun to shine on it.

In the spring, the pond should be filled again and the fences removed. Some people just fence the pond in the winter, dig a new pond and fill in the old one with new pond's dirt.

Mycoplasma mastitis is often cultured in water from cooling ponds. This is a real problem if pulsators do not work well resulting in damaged teat ends. Many herds with mycoplasma positive ponds, have no mycoplasma in the bulk tank.

If the water is filthy, cows should not be permitted in the ponds and the ponds should be pumped dry. The sludge should be cleaned and the ponds filled with water again. Both the sludge and the water shoud be removed from cow areas.

Cooling Ponds to Reduce Body Temperature

In the summer of 1986, an experiment was conducted at a North Florida dairy that was using man-made cooling ponds. Ten early lactation Holstein cows were fitted with radio transmitters that transmitted inner ear temperatures every five minutes to a radio receiver and data logger.

Data were collected over a four-day period in mid-August. The cows' activities were also monitored during this time. Entering and exiting the ponds, eating, drinking and laying were recorded.

The cooling ponds lowered the cows' temperature by 1-2°F depending on the time of day they entered the cooling ponds. The average length per stay in the pond was 18 minutes for events from midnight to noon and 12 minutes per visit from noon to midnight. It was obvious that cooling ponds cooled cows.

Pond Water Quality

In 1987, samples were taken from a dairy's man-made earthen ponds. The ponds were sampled weekly from May 19,1987 to September 28, 1987. Six to ten ponds were sampled each week.

Total bacteria counts (TBC), varied greatly from week to week. This may have been due to no water entering the ponds. There did not seem to be any great increase in total bacteria counts or coliform counts as time progressed. The total bacteria count averaged 3,133,700 CFU/ml for all ponds for the 20 weeks. The coliform counts averaged 14,340 CFU/ml for the same period.

Cooling Ponds and Their Effect on Milk Quality

In a previous study on a dairy in Florida with man-made cooling ponds, it was reported that cows exposed to cooling ponds did not experience more clinical mastitis than cows that did not have access to cooling ponds. In fact, cows exposed to cooling ponds during the trial period were only half as likely to develop a case of clinical mastitis as cows not exposed to cooling ponds.

Clinical Mastitis Organisms

The same dairy had provided its clinical mastitis records for several years. The total number of clinical mastitis cases was high in the first quarter of the year and then declined sharply about the time hot weather came and the cows started using the ponds. The incidence stayed low, even in the last quarter of the year when the ponds were not in use. This can be explained many ways. The first quarter of the year was an extremely wet period.

The cows were quite dirty and the cow wash system was available to one-half of the herd. This may explain some of the variation, even though one-half of the herd had never had a cow wash. Another confounding variable was that this herd switched to Clorox for pre- post-teat dipping in April of that year. It had previously pre- and post-dipped with a Chlorhexidine product. From this it could be concluded that Clorox caused the reduction of clinical mastitis. However, there were no controls in this study, one could conclude nothing except that the ponds did not increase clinical mastitis.

This herd was on a lactation treatment study during the summer of 1987. A total of 40 cows were sampled and treated. There were no unusual organisms treated. These 40 cows were not all the cows treated for clinical mastitis during the period the cows had access to the ponds, but the results were encouraging.

Statewide Effects of Ponds on Milk Quality

Regulatory samples for all the herds in Florida were obtained from the Division of Dairy Industry for the year of 1987 and were analysed for the effects of no ponds, man-made ponds and natural ponds. The numbers presented are least squares estimates of Standard Plate Counts (SPC) and Direct Microscope Somatic Cell Counts (DMSCC) for each month. The data are not biased by different numbers of dairies with no ponds, natural, or man-made ponds. The data were analysed by herds whose milking cows had access to ponds as indicated on the survey.

The data indicated that milking cows with access to man-made cooling ponds had lower SPC and DMSCC counts than other herds in the state with no access to ponds and those herds with natural ponds were higher in both categories. From these data it could be argued that people who built man-made ponds were better managers and thus had higher milk quality.

FACILITY DESIGN AND MANAGEMENT OF ANIMALS

The effect of animal facility design and management can have a major impact on all types of emissions. Specific research that has investigated these factors has generally determined large variations in airborne emissions of contaminants like ammonia or dust. Unfortunately, all of the management factors and environmental conditions contributing to these changes in emissions are not well understood or documented.

It has been shown that odour and gaseous emissions from buildings are increased if the walls and floors are constantly covered with layers of feces and urine. Design modifications are based on reducing the area of the emitting surfaces, frequent removal of slurry from the houses, movement of slurry through slats, temperature control and ventilation rates. Use of sloped "catch pans", gutters and narrow collection channels help reduce emitting surfaces under the slats. Reductions in ammonia emission from new buildings varied from 30 to 70 per cent as compared to conventional buildings.

In the United States, hoop structures with straw bedding are being considered as an alternative to large-scale confinement structures for swine production. On deep litter systems (6.8 kg straw pig^{-1} day^{-1}), ammonia emission is comparable with emission from a fully slatted floor barn. Emissions can be kept at low levels by increasing the amount of straw or by allowing partial urine drainage. However, emissions of nitrogen gases in deep litter systems tend to be higher due to the formation of N_2O which contributes to the greenhouse effect and affects the ozone layer.

Traditional methods of NH_3 control in buildings have involved removal of manure, drying of manure to avoid or reduce urease breakdown, and litter amendments to control pH in broiler litter. Groot Koerkamp *et al*. reported on the effects of a litter drying system on the composition of the litter and the emission of ammonia from a tiered wire floor poultry housing system for layers. They concluded that forced air movement (0.5 m^3 hr^{-1} per hen) above the litter enhanced the evaporation of water from the litter substantially as compared to no forced air movement above the litter. Litter dry matter content was kept above 900 g kg^{-1} and the Total Ammoniacal Nitrogen (TAN) concentration (0.7 g kg^{-1}) and pH (7.3) decreased as compared to the composition of litter in poultry houses without drying of litter. The change in litter composition apparently helped lower ammonia emissions. The lowest levels of ammonia emission (about 2.0 mg hen^{-1} hr^{-1}) were recorded when manure was removed more frequently and more ventilation was provided.

Yang *et al*. determined nitrogen losses from four high-rise laying hen houses located in Iowa. Nitrogen losses were between 25 and 41 per cent based on Total Kjeldahl Nitrogen (TKN) in feed. They found that the higher the moisture content of manure, the higher the ratio between NH_3 and TKN in manure, and therefore, the higher the percentage of N loss. These findings are in reasonable agreement to the conclusions reached by Dutch researchers in a previous study described above. It is a common broiler industry practice to manipulate minimum ventilation rates continuously to strike a balance between the need for energy conservation (supplemental heat must be provided during cold weather) and indoor air quality. Recent advances in water delivery systems have greatly improved poultry environments, to the point where problems with dust and gases have replaced humidity as a common complaint to extension personnel and consultants. With a tendency for lower litter moisture content, less ammonia is generated and volatilized. However, this may be offset by an industry practice of reusing broiler litter for multiple flocks; if litter moisture is high enough to support urease breakdown then the potential for high ammonia emission exists because total litter N is greater. Ammonia emissions from cattle housing is usually influenced by the flooring system, type of bedding and manure handling system (slats, scrape, or flush). Kroodsma *et al.* determined the effects of different floor types and flushing on ammonia emission rates from free-stall dairies. Scraped or dirty solid floors gave the highest ammonia emission (about 15 g NH_3 m^{-2} day^{-1}), while flushing gave the lowest (5 g NH_3 m^{-2} day^{-1}). Scraped or dirty slatted floors were found to emit about 9 g g NH_3 m^{-2} day^{-1}.

Braam *et al.* looked at practical ways to reduce ammonia emission from double-sloped solid floors with a central urine gutter in dairy houses. They found that ammonia emission from the compartment with the double-sloped solid floor operating with one urine gutter and without water spraying was reduced by 50 per cent when compared to a control (slatted floor with underfloor slurry pit). Ammonia emission was further reduced when water was sprayed after scraping. Swierstra *et al.* have recently reported on a grooved floor system consisting of prefabricated concrete elements with perforations spaced 1.1 m apart to channel urine from the floor. Feces were removed every two hours by a mechanical scraper and were dumped into the pit through a floor opening at the end of the alley.

The blade of the scraper was equipped with a tooth-shaped rubber strip to clean the grooves. Ammonia emissions from the grooved floor were found to be 46 per cent less than emissions from a reference floor (concrete slotted floor). Closing of the perforations resulted in an ammonia emission reduction of only 35 per cent compared to the reference floor. Demmers *et al.* showed that ammonia emission from straw bedded beef housing was 40 per cent less than ammonia emission from a slurry-based dairy unit. Jeppsson concluded that ammonia emission from deep-bedded housing for heifers using a mixture of peat (60 per cent) and chopped straw (40 per cent) was reduced by almost 60 per cent as compared to bedded areas with long straw. Reduction of ammonia emission was attributed to the ability of peat to absorb water and ammonia, lower pH level, and also to its high C/N ratio. Ammonia emissions were 8 and 18 g NH_3/m^2-day, for the peat-straw mixture and long and chopped straw bedding, respectively. In addition, ammonia emission from the manure alley was found to be significantly less than from the bedding area with straw bedding.

Hydrogen Sulfide

Hydrogen sulfide is formed by bacterial sulfate reduction and the decomposition of sulfur-containing organic compounds in manure under anaerobic conditions. H_2S gas is colorless, heavier than air, highly soluble in water and has the characteristic odour of rotten eggs at low concentrations. At concentrations around 30 ppb the H_2S odour can be detected by over 80 per cent of the population. The U.S. OSHA has implemented a 10 ppm limit for indoor 8-hour H_2S exposures to protect human worker health. Most human health problems associated with hydrogen sulfide emissions are related to emissions from paper mills, refineries, and meat packing plants. Currently, there is only circumstantial evidence relating emission of hydrogen sulfide from livestock and poultry to human health.

Although there are health risks associated with high concentrations of H_2S, concentrations are usually very low in and around animal housing as compared to concentrations of CO_2 and NH_3. Ni *et al.* and Ni *et al.* measured H_2S concentrations between 65 and 536 ppb in swine finishing facilities in Indiana. Bicudo *et al.* measured hydrogen sulfide concentrations continuously during 30-day periods around swine buildings in Minnesota. A maximum of 450 ppb of H_2S was recorded at 5 m downwind from a naturally ventilated finishing barn.

Mean H_2S concentrations around a nursery (mechanically ventilated) and wean-to-finish (naturally ventilated) barns were between 4.5 and 10.9 ($\pm$0.3) ppb. H_2S levels around a hoop barn were lower than 2 ppb. Zhu *et al.* (2000b) studied the daily variations in H_2S emissions from various mechanically and naturally ventilated swine housing systems in Minnesota. H_2S concentrations varying between 200 and 3,400 ppb were reported.

Koelsch *et al.* measured total reduced sulfur levels in a beef cattle feedlot using a Jerome meter. This instrument measures total reduced sulfur (TRS) compounds, including alkyl sulfides, disulfides, mercaptans, and cyclic sulfur compounds. Concentrations in the center of the feedlot varied between 1 and 14 ppb. Clark and McQuitty recorded a maximum H_2S level of 145 ppb in four of six commercial free-stall dairy barns in Alberta. McQuitty *et al.* reported on H_2S concentrations in three commercial laying barns under winter conditions. No detectable traces of H_2S were found in two barns and a maximum H_2S concentration of 30 ppb was measured in the third barn. Several researchers have studied the effects of swine dietary sulfur intake on H_2S levels in pig housing. Shurson *et al.* reported a reduction in H_2S emissions from nursery pigs fed a low sulfur diet as compared to a traditional diet. Donham *et al.* documented a positive, but not significant, correlation between sulfate levels in drinking and cleaning water and the sulfide content in swine manure. A slightly positive relationship between total sulfides in manure and hydrogen sulfide concentration in the building exhaust air was also reported.

A limited amount of research has focused on H_2S emissions from animal housing. Most of this data is from swine facilities. Measurements obtained by Zhu *et al.* were reported for a 12-hour period, and values shown in Table were not converted to a 24-hour period.

Table. Hydrogen Sulfide Emission Factors from Livestock and Poultry Housing

Species	Production unit	Notes	Emission Factor ($g H_2S AU^{-1} day^{-1}$)	Reference
Pig	Finish	Fully slatted	2.4-22.6	Ni et al. (2002)
	Finish	Fully slatted – no pigs	0.22-0.49	Ni et al. (2000)
	Finish	Fully slatted	1.25	Ni et al. (2000)
	Finish	Fully slatted (mechanically ventilated)	5	Zhu et al. (2000a)
	Finish	Fully slatted (naturally ventilated)	2-7	Zhu et al. (2000a)
	Farrowing	Fully slatted	4	Zhu et al. (2000a)
	Gestation	Fully slatted	1	Zhu et al. (2000a)
	Nursery	Fully slatted	23-160	Zhu et al. (2000a)
Poultry	Broiler	Litter	3.3	Zhu et al. (2000a)

Hydrogen sulfide emissions from swine and poultry housing tend to be under 5 g H_2S AU^{-1} day^{-1}. Ni *et al.* found that diurnal fluctuations and differences between daily H_2S mean concentrations were relatively large and that spatial differences were not significant when averaged over long durations. Gay *et al.* reported on H_2S emissions rates from 80 farms in Minnesota. Mean H_2S emissions varied from 0.02 to 1.5 g H_2S m^{-2} day^{-1} from swine housing, from 0.03 to 0.35 g H_2S m^{-2} day^{-1} from poultry housing, from 0.09 to 0.25 g H_2S m^{-2} day^{-1} from dairy housing, and were about 0.15 g H_2S m^{-2} day^{-1} from beef feedlots. Ventilation rates were measured as explained before.

This data set was subject to large variability and it is difficult to compare it to other data reported in terms of AU. However, this data indicates that H_2S emissions are consistently higher for swine housing as compared to poultry and dairy housing and beef feedlots. More data is needed to identify baseline H_2S emissions from livestock and poultry housing.

Methane

Methane (CH_4) is produced by the microbial degradation of soluble lipids, carbohydrates, organic acids, proteins, and other organic components. CH_4 is another strong greenhouse gas.

Table. Estimated Methane Emissions from Livestock and Poultry Waste

Animal Type	CH_4 Emission Factor (kg CH_4 animal-1 year)
Cattle in feedlots	23
Dairy	70
Swine	20
Caged Layer	0.3
Broiler	0.09

The presence of atmospheric CH_4 has been associated with climatic changes: Sommer and Moller reported that CH_4 contributes between 9 and 20 per cent to the total global warming potential. The estimated of CH_4 contributions from various livestock and poultry species. These CH_4 emission estimates were based on standard methane conversion factors (MCF) applied to a global scale rather than actual measurements.

The MCFs used by Safley and Casada were based on manure handling method, temperature, and the amount of volatile solids in manure. Steed and Hashimoto conducted a laboratory experiment to verify the estimated MCF values for dairy cows used by Safley and Casada. This research indicated that the MCF was less for dry manure under aerobic conditions, such as that found on feedlots and pastures, than for liquid or solid manure storage systems.

Table. Measured Methane Emission Factors (MCF) for Dairy Cows

System Type	**MCF estimates by Safley and Casada (1992)**	**MCF measured at 20°C Steed and Hashimoto (1994)**
Pasture/Feedlot	10	0.3
Liquid slurry	20-90	55.3
Solid	10	45.7

Kaharabata and Schuepp used an atmospheric tracer (SF_6) to estimate CH_4 emissions from dairy cattle housed in a barn and feedlot. The tracer gas was released from sixteen point sources distributed within the barn or feedlot to simulate the CH_4 release from cows. Predicted CH_4 emissions from the barn and feedlot were 542 L CH_4 cow day and 631 L CH_4 cow^{-1} day^{-1}, respectively. Overall uncertainty of the results was approximately 30 per cent. Osada *et al.* measured CH_4 emissions from an experimental swine finishing unit with a slatted floor during an 8-week period. Methane emissions varied from 48 to 54 g CH_4 AU^{-1} day^{-1}. Zahn *et al.* measured CH_4 emissions from deep-pit and pull-plug swine finishing facilities during August and September of 1997. Methane emissions of 160 g CH_4 AU^{-1} day^{-1} were reported.

Non-Methane Volatile Organic Compound

Animal housing and manure handling systems generate a variety of gases. Most of the research conducted to date has not quantified VOC emissions but rather documented the generation of these gases. Kreis developed one of the earliest lists of volatile compounds associated with decomposition of cattle, poultry, and swine wastes. He listed 32 compounds reported to have come from cattle wastes, 17 from poultry wastes, and more than 50 compounds from swine wastes. Hartung and Phillips reported quantitative information on concentrations found in the air of animal houses for 23 VOC. O'Neill and Phillips compiled a list of 168 different compounds identified in swine and poultry wastes. More recently, Schiffman *et al.* identified a total of 331 different VOC and fixed gases from swine facilities in North Carolina. These odourous compounds are usually produced and accumulated in collection and storage systems where feces and urine are decomposed by bacteria under anaerobic conditions. There are four different chemical classes of VOC: volatile fatty acids (VFA), indoles and phenols, amines, and sulfur-containing compounds. The VFA group consists of acetic, propionic, butyric, iso-butyric, valeric, iso-valeric, caproic, and capric acids. Indole, skatole, cresol, 4-ethylphenol appear to be the major odourants included in the indole and phenol group.

Phenolic compounds are produced from the microbial degradation of amino-acids such as tyrosine in the intestinal tract of animals. Volatile amines include compounds such as methylamine, ethylamine, putrescine, etc. The main components of the sulfur-containing group are sulfides as well as methyl- and ethyl- mercaptans. These compounds are produced by the reduction of sulfate and by bacterial degradation of sulfur-containing amino-acids. Zhu provided a thorough review of the microflora in swine manure and its potential to produce odourous volatile compounds. Information on VOC emissions from animal housing is limited. Zahn *et al.* measured VOC emissions from pull-plug and deep-pit swine houses during August and September 1997. Twelve different non-methane VOCs were detected at a total concentration of 806 mg m^{-3}. The VOC mixture consisted primarily of acetic, propionic, and butyric acid. Estimated VOC emissions were 90 g VOC $hour^{-1}$.

Dust

Particulates in and around animal production sites include soil particles, bits of feed, dried skin, hair or feathers, dried feces, bacteria, fungi, and endotoxins. Sources include animals, feed storage and processing sites, floors, manure storage and handling equipment, open lots, compost sites, and other elements of animal agriculture systems. Feed was found to be the primary component of the dust in animal housing. Soil particles from open unpaved feedlots also contribute to dust levels. Dust emissions from feedlots depend on soil texture, rainfall, feedlot surface moisture content, wind speed, season, and other factors. The white paper on particulate matter emissions from confined animal feeding operations – management and control measures provides more specific information on dust emission from cattle feedlots.

Flooring design has been shown to significantly affect the airborne dust levels; solid floors have much higher levels than open-mesh floors. The latter allow feces and soiled bedding to fall below the floor level and minimize dust generated by animal activities. There is little research on dust emission factors from animal agriculture facilities and their environmental impact. Most studies have focused on dust concentrations and characterization in swine and poultry housing rather than emissions. Limited information is available on dust concentrations in dairy and horse facilities. Auvermann *et al.* summarize information on particulate matter in swine and poultry housing as well as on open cattle feedlots. Other studies have concentrated on the effects of dust in confinement housing on human worker and animal health. Impacts of particulate matter and bioaerosols on human health are discussed in detail in the white paper on health effects of aerial emissions from animal production and waste management systems.

Diet manipulation of Feeding Management

Use of improved feeding management practices, selective feed ingredient use, precision in diet formulation, and dietary electrolyte balance has been shown to reduce nutrient excretion, and subsequent, odour and gas emissions from livestock manure. *Yucca schidigera* extract has been shown to reduce ammonia emission from manure by inhibiting urease activity. Sutton *et al.* showed that ammonia emission was suppressed by 55.5 per cent in swine manure from pigs fed sarsaponin extract at a rate of 4 oz/ton of feed, but Kemme *et al* was unable to verify this response, and showed that much higher amounts of the extract (6,000 ppm) was needed for maximal suppression of ammonia from urea.

Reduced crude protein diets containing synthetic amino acids have been shown to reduce nitrogen excretion in pigs, which can lead to potential reduced ammonia emissions. Reductions in ammonia emissions from 28 to 79 per cent through diet modifications in swine have been reported. Ferguson *et al.* have examined the effects of diet manipulation on the litter equilibrium NH_3 gas concentration in broiler housing. Gas was sampled using an equilibrium chamber. Equilibrium concentrations between 53 and 83 ppm were obtained for different diet treatments. Reducing crude protein caused equilibrium NH_3 gas concentration to decline by about 30 per cent. Gates *et al.* also reported on the effect of reduced crude protein on equilibrium NH_3 broiler litter. Equilibrium NH_3 concentrations varied from 0 to 161 ppm, depending on the flock number, ventilation rate, and diet treatment. A low crude protein diet resulted in about 90 per cent reduction in equilibrium NH_3 concentration even for used litter. The differences between the Gates and Ferguson studies were basically litter moisture content and number of flocks. Gates worked with significantly drier litter (16 – 25 per cent) than Ferguson (50 – 60 per cent) and took measurements over a period equivalent to the raising of three flocks using the same litter, while Ferguson data is from one flock only. Reduced crude protein diets also help reduce NH_3 emissions from dairy and beef cattle. James *et al.* reported a 28 per cent reduction in NH_3 emission from dairy cows fed a low crude protein ration.

Smits *et al.* observed further reductions in NH_3 emissions as compared to James *et al.* study. Klopfenstein and Erickson observed reductions in NH_3 emissions from the surface of beef cattle feedlots between 15 and 30 per cent when cattle were fed a lower crude protein diet. Diet manipulation as well as its effects on manure production and composition is addressed in detail in another white paper.

Nitrous Oxide

Nitrous oxide is a product of both nitrification and denitrification. Pahl *et al.* (2001) demonstrated that there was a large variation in the split between nitrification and denitrification processes as the source of N_2O production. Their results showed that specific conditions could favour nitrification or denitrification to be the principal source of N_2O emissions: (i) through denitrification under oxygen inhibition; or (ii) through nitrification in aerobic systems, in combination with the presence of nitrification products. Therefore, N_2O can be released at any stage of livestock production where conditions favour these processes. Leaching, absorption by plants, or utilization by microorganisms indirectly influences the production of N_2O. Nitrous oxide emissions are an environmental concern. Houghton *et al.* stated that N_2O is approximately 200 times more efficient than CO_2 in absorbing infrared radiation. Methane, another strong greenhouse gas, is only 26 times more efficient than CO_2 in absorbing infrared radiation. Furthermore, N_2O contributes to the reduction of ozone in the stratosphere through the photochemical decomposition of N_2O to NO.

Data on N_2O emissions from animal housing is limited. Osada *et al.* measured N_2O emissions from an experimental swine finishing unit with a slatted floor during an 8-week period. Nitrous oxide emissions varied from 0.8 to 2.1 g N_2O AU^{-1} day^{-1}. Emissions were reduced when underground manure pits were discharged weekly. Chadwick *et al.* summarized N_2O emissions from animal housing in the U.K. Nitrous oxide emissions varied from 0.4 to 26 g N_2O AU^{-1} day^{-1}. The lowest emissions values were from swine housing and the highest were from poultry housing. Chadwick *et al.* also noted that dairy housing with slurry-based systems had significantly lower N_2O emissions than dairy housing that used straw bedding.

Hazardous Component of Endotoxin

Endotoxin is a hazardous component of airborne particulates in animal operations. It arises from the degradation of Gram-negative bacterial cell wall and is ubiquitous in the agricultural environment. Endotoxin is a potent inflammatory agent that produces systemic effects and lung obstruction, even at low levels of exposure. Despite a clear recognition that inhaled endotoxin is an occupational hazard in livestock and poultry confinement housing, cotton processing, vegetable processing, fibreglass manufacturing, and metal machining environments, there are no established occupational exposure limits in the United States or Canada. This is probably due to the fact that endotoxin exposure assessment methods have not been adequately optimized and validated. Wathes *et al.* measured endotoxin emissions from broiler and layer facilities in the U.K. Endotoxin emissions varied between less than 1 and 10 g AU^{-1} h^{-1} in the winter, and between 30 and 45 g AU^{-1} h^{-1} in the summer.

2

Nutritional Requirements of Higher Animals

Cells use nutrients as fuel for energy production (catabolism) and as material for processes of maintenance and growth (anabolism). Multicellular animals derive energy solely from the breakdown of complex organic molecules, mainly carbohydrates and fats. Because the fuel for the maintenance of animal life comes only from other living organisms or their remains, animals are known as heterotrophic organisms. All animal life depends ultimately on the existence of organisms (largely green plants) that can use inorganic sources of energy, of which solar radiation is by far the most important; some microorganisms, however, obtain energy from oxidation of simple inorganic compounds.

For anabolic purposes, food must provide adequate amounts of all chemical elements needed by the cells. Of the approximately 35 elements now known to occur in animal cells, four (oxygen, carbon, hydrogen, and nitrogen) make up about 95 per cent of the cell weight; another nine (calcium, phosphorus, chlorine, sulfur, potassium, sodium, magnesium, iodine, and iron) contribute about 4 per cent. All of these elements have indispensable functions. The remaining 20-odd, together constituting less than 1 per cent of cell weight, are called trace elements, because they occur in minute quantities. Although some of them may become incorporated into cells by accident, many fulfill vital functions.

It is important to note that animal cells cannot synthesize from simple compounds certain necessary complex molecules. Instead, certain large organic molecules must serve as building blocks; such so-called essential dietary components include the vitamins, some amino acids, and certain fatty substances. In general, higher animals appear to have more restricted synthetic powers than lower ones and to require a correspondingly greater number of essential foodstuffs. Microorganisms in the intestines of vertebrates may synthesize materials essential for the host, so that the food of the latter need not contain these substances.

Buffaloes are strict grazers and only browse when feed is utterly scarce. Normally, buffaloes graze during the day. In case of extremely high ambient temperatures, grazing takes place in the morning and afternoon and sometimes during night time. Buffaloes graze more and better than cattle. Thereby they consume more feed and nutrients per kg of body weight than cattle do.

Newborn calves suckle their mothers within two hours of birth. Normal suckling frequency is approximately 6 to 8 times per day. The calves start to nibble grass at 3 to 4 weeks of age, although they are not actually grazing until after a few weeks more. When the calves have reached two months of age, forage starts to become more important than previously and soon most of the nutrient intake comes from forage rather than milk. Natural weaning of calves is usually within a year or before its mother's next parturition.

NUTRITION FOR HORSE

Knowledge of horse nutrition has grown by leaps and bounds during the last 15 years. Research has become more precise and critically evaluated. But more important, this research has given horse owners greater understanding of nutrition. They are more aware of the basic nutrients required by all classes of horses, than in past years. Equine nutrition refers to the feeding of horses, ponies, mules, donkeys, and other equines. Correct and balanced nutrition is a critical component of proper horse care. Horses are herbivores, a type of non-ruminant known as a "hind-gut fermentor." What this means is that horses have only one stomach, similar to humans. However, unlike humans, they also have to digest plant fibre that comes from grass and hay. Therefore, unlike ruminants, who digest fibre in plant matter by use of a multichambered stomach, horses use bacterial fermentation that occurs in the organ known as the *cecum* (or *caecum*) to break down cellulose.

Fig. Grass is a Natural Source of Nutrition for a Horse

In practical terms, horses prefer to eat small amounts of food steadily throughout the day, as they do in nature when grazing on pasture. Although this is not always possible with modern stabling practices and human schedules that favour feeding horses twice a day, it is important to remember the underlying biology of the animal when determining what to feed, how often, and in what quantities. The digestive system of the horse is somewhat delicate. Because horses are unable to regurgitate food, except from the esophagus, if they overeat or eat something poisonous, vomiting is not an option. They also have a long, complex large intestine and a balance of beneficial bacteria in their cecum that can be upset by rapid changes in feed. Because of these factors, they are very susceptible to colic, which is a leading cause of death in horses. Therefore, horses require clean, high-quality feed, provided at regular intervals, and may become ill if subjected to abrupt changes in their diets. Horses are also sensitive to molds and toxins. For this reason, they must never be fed contaminated fermentable materials such as lawn clippings. Fermented silage, sometimes called "haylage" is fed to horses in some places; however, contamination or failure of the fermentation process that allows any mold or spoilage may be toxic. Horses and other members of the *Equidae* family are adapted by evolutionary biology to eating small amounts of the same kind of food all day long. In the wild, horses ate prairie grasses in semi-arid regions and traveled significant distances each day in order to obtain adequate nutrition. Therefore, their digestive system was made to work best with a small but steady flow of food that does not change much from day to day. Digestion begins in the mouth. First, the animal selects pieces of forage and picks up finer foods, such as grain, with sensitive, prehensile, lips.

The front teeth of the horse, called incisors, clip forage, and food is ground up for swallowing by the premolars and molars. The esophagus carries food to the stomach. The esophagus enters the stomach at an acute angle, creating a one-way valve, with a powerful spincter mechanism at the gastroesophageal junction, which is why horses cannot vomit. The esophagus is also the area of the digestive tract where horses may suffer from choke. Horses have a relatively small stomach for their size, which limits the amount of feed that can be taken in at one time. The average sized horse has a stomach with a capacity of only four gallons, and works best when it contains about two gallons.

One reason continuous foraging or several small feedings per day are better than one or two large meals is because the stomach begins to empty when it is 2/3 full, whether the food in the stomach is processed or not. The small intestine is 50 to 70 feet long and holds 10 to 12 gallons. This is the major digestive organ where most nutrients are absorbed. Bile from the liver acts here, combined with enzymes from the pancreas and small intestine itself. Equids do not have a gall bladder, so bile flows constantly, another reason for a slow but steady supply of food. Most nutrition is absorbed into the bloodstream from the small intestine. The cecum is the first section of the large intestine. It is also known as the "water gut" or "hind gut." It is a cul-de-sac pouch, about 4 feet long that holds 7 to 8 gallons. It digests cellulose plant fibre through bacterial fermentation. The reason horses must have their diets changed slowly is so the bacteria in the cecum are able to modify and adapt to the different chemical structure of new feedstuffs. Too abrupt a change in diet can cause colic, as the new food is not properly digested.

The large colon, small colon, and rectum make up the remainder of the large intestine. The large colon is 10-12 feet long and holds up to 20 gallons of semi-liquid matter. Its main purpose it to absorb carbohydrates which were broken down from cellulose in the cecum. Due to its many twists and turns, it is a common place for a type of horse colic called an impaction. The small colon is 10-12 feet long, holds about 5 gallons, is the area where the majority of water is absorbed, and where fecal balls are formed. The rectum is about one foot long, and acts as a holding chamber for waste, which is then expelled from the body via the anus.

Nutrients

Like all animals, equines require five main classes of nutrients to survive: water, energy (primarily in the form of fats and carbohydrates), proteins, vitamins, and minerals. Water makes up between 62-68 per cent of a horse's body weight and is essential for life. Horses can only live a few days without water, becoming dangerously dehydrated if they lose 8-10 per cent of their natural body water. Therefore, it is critically important to provide access to a fresh, clean, and adequate supply of water. An average 450kg horse drinks 10 to 12 gallons of water per day, more in hot weather, when eating dry forage such as hay, or when consuming high levels of salt, potassium, and magnesium.

Horses drink less water in cool weather or when on lush pasture, which has a relatively higher water content. When under hard work, or if a mare is lactating, water requirements may be as much as four times greater than normal. Though they need a great deal of water, horses spend very little time drinking; usually 1-8 minutes a day, spread out in 2-8 episodes. Water plays an important part in digestion. The forages and grains horses eat are mixed with saliva in the mouth to make a moist bolus that can be easily swallowed. Therefore, horses produce up to 10 gallons of saliva per day. Nutritional sources of energy are fat and carbohydrates. Protein is a critical building block for muscles and other tissues. Horses that are heavily exercised, growing, pregnant or lactating need increased energy and protein in their diet. However, if a horse has too much energy in its diet and not enough exercise, it can become too high-spirited and difficult to handle.Fat exists in low levels in plants and can be added to increase the energy density of the diet.

Fat has 9 Mcal/kg of energy, which is 2.25 times that of any carbohydrate source. Because equids have no gall bladder to store large quantities of bile, which flows continuously from the liver directly into the small intestine, fat, though a necessary nutrient, is difficult for them to digest and utilize in large quantities. However, they are able digest a greater amount of fat than can cattle. Horses benefit from up to 8 per cent fat in their diets, but more does not always provide a visible benefit. Horses can only have 15 per cent-20 per cent fat in their diet without the risk of developing diarrhea. Carbohydrates, the main energy source in most rations, are usually fed in the form of hay, grass and grain. Soluble carbohydrates such as starches and sugars are readily broken down to glucose in the small intestine and absorbed. Insoluble carbohydrates, such as fibre (cellulose), are not digested by enzymes, but are fermented by microbes in the cecum and large colon to break down and release their energy sources, the volatile fatty acids.

Soluble carbohydrates are found in nearly every feed source; corn has the highest amount, then barley and oats. Forages normally have only 6 to 8 per cent soluble carbohydrate, but under certain conditions can have up to 30 per cent. Sudden ingestion of large amounts of starch or high sugar feeds can cause colic or laminitis.Protein is used in all parts of the body, especially muscle, blood, hormones, hooves, and hair cells. The main building blocks of protein are amino acids. Alfalfa and other legumes in hay are good sources of protein that can be easily added to the diet. Most adult horses only require 8 to 10 per cent protein in their diet; however, higher protein is important for lactating mares and young growing foals. Horses that are not subjected to hard work or extreme conditions usually have more than adequate amounts of vitamins in their diet if they are receiving fresh, green, leafy forages. Sometimes a vitamin supplement is needed when feeding low-quality hay, if a horse is under stress (illness, traveling, showing, racing, and so on), or not eating well. Grain has a different balance of nutrients than forage, and so requires specialized supplementation to prevent an imbalance of vitamins and minerals.

Minerals are required for maintenance and function of the skeleton, nerves and muscles. These include calcium, phosphorus, sodium, potassium, and chloride, and are commonly found in most good-quality feeds. Horses also need trace minerals such as magnesium, selenium, copper, zinc and iodine. Normally, if adult animals at maintenance levels are consuming fresh hay or are on pasture, they will receive adequate amounts of minerals in their diet, with the exception of sodium chloride (salt), which needs to be provided, preferably free choice. Some pastures are deficient in certain minerals, especially selenium or copper, and in such situations, deficiency diseases may occur if the horses' feed is not supplemented. Calcium and phosphorus are needed in a specific ratio of between 1:1 and 2:1. Adult horses can tolerate up to a 5:1 ratio, foals no more than 3:1. A total ration with a higher ratio of phosphorus than calcium is to be avoided. Over time, the imbalance will ultimately lead to bone problems such as osteoporosis. Foals and young growing horses through their first three to four years have special nutritional needs and require feeds that are balanced with a proper calcium:phosphorus ratio and other trace minerals. A number of skeletal problems may occur in young animals with an unbalanced diet. Hard work increases the need for minerals; sweating depletes sodium, potassium, and chloride from the horse's system. Therefore, supplementation with electrolytes may be required for horses in intense training, especially in hot weather.

TYPES OF FEED

Equids can consume approximately 2 per cent to 2.5 per cent of their body weight in dry feed each day. Therefore, a 1000 lb adult horse could eat up to 25 pounds of food. Foals less than six months of age eat 2 to 4 per cent of their weight each day.Solid feeds are placed into three categories: forages, concentrates (including grain or pelleted rations), and supplements. Equine nutritionists recommend that 50 per cent or more of the animal's diet by weight should be forages. If a horse is working hard and requires more energy, the use of grain is increased and the percentage of forage decreased so that the horse obtains the energy content it needs for the work it is performing. However, forage amount should never go below 1 per cent of the horse's body weight per day.

Forages, also known as "roughage," are plant materials classified as legumes or grasses, found in pastures or in hay. Often, pastures and hayfields will contain a bland of both grasses and legumes. Nutrients available in forage vary greatly with maturity of the grasses, fertilization, management, and environmental conditions. Grasses are tolerant of a wide range of conditions and contain most necessary nutrients. Some commonly used grasses include timothy, brome, fescue, coastal bermuda, orchard grass, and Kentucky bluegrass. Legumes such as clover or alfalfa are usually higher in protein, calcium, and energy than grasses.

However, they require warm weather and good soil to produce the best nutrients. Legume hays are generally higher in protein than the grass hays. They are also higher in minerals, particularly calcium, but have an incorrect ratio of calcium to phosphorus. Because they are high in protein, they are very desirable for growing horses or those subjected to very hard work, but the calcium:phosphorus ratio must be balanced by other feeds to prevent bone abnormalities. Hay is a dried mixture of grasses and legumes. It cut in the field and then dried and baled for storage. Hay is most nutritious when it is cut early on, before the seed heads are fully mature and before the stems of the plants become tough and thick.

Hay that is very green can be a good indicator of the amount of nutrients in the hay; however, colour should not be used as sole indicator - smell and texture are also important. Hay can be analyzed by many laboratories and that is the most reliable way to tell the nutritional values it contains. Moldy or dusty hay should not be fed to horses, as it is the most common cause of Recurrent airway obstruction, also known as COPD or "heaves." Hay, particularly alfalfa, is sometimes compressed into pellets or cubes. Processed hay can be of more consistent quality and is more convenient to ship and to store. It is also easily obtained in areas that may be suffering localized hay shortages. However, these more concentrated forms can be overfed and horses are somewhat more prone to choke on them. On the other hand, hay pellets and cubes can be soaked until they break apart into a pulp or thick slurry, and in this state are a very useful source of food for horses with tooth problems such as dental disease, tooth loss due to age, or structural anomalies.

Haylage, also known as *Round bale silage* is a term for grass sealed in airtight plastic bags, a form of forage that is frequently fed in the United Kingdom and continental Europe, but is not often seen in the United States. Because haylage is a type of silage, hay stored in this fashion must remain completely sealed in plastic, as any holes or tears can stop the preservation properties of fermentation and lead to mold or spoilage. Rodents chewing through the plastic can also spoil the hay introducing contamination to the bale.If a rodent dies inside the plastic, the subsequent botulism toxins released can contaminate the entire bale. Another type of forage sometimes provided to horses is beet pulp, a byproduct left over from the processing of sugar beets, which is high in energy as well as fibre.Sometimes, straw or chaff from cereal grains is fed to animals. However, this is roughage with little nutritional value. It is not recommended by nutritionists as a horse feed, though it is sometimes used as a filler; it can slow down horses who eat their grain too fast, or it can provide additional fibre when the horse must meet most nutritional needs via concentrated feeds. Straw is more often used as a bedding in stalls to absorb wastes.

Whole or crushed grains are the most common form of concentrated feed, sometimes referred to generically as "oats" or "corn" even if those grains are not present, also sometimes called "straights" in the UK. Oats are the most popular grain for horses. Oats have a lower digestible energy value and higher fibre content than most other grains. They form a loose mass in the stomach that is well suited to the equine digestive system. They are also more palatable and digestible other grains. Corn, referred to as Maize in the UK, is the second most palatable grain. It provides twice as much digestible energy as an equal volume of oats and is low in fibre. Because of these characteristics, is easy to over-feed corn, causing obesity, so horses are seldom fed corn all by itself. Nutritionists caution horse owners that moldy corn should never be fed because it is poisonous to horses. Barley is also fed to horses, but needs to be processed to crack the seed hull and allow easier digestibility. It is frequently fed in combination with oats and corn, a mix informally referred to by the acronym "COB" (for Corn, Oats and Barley).

Wheat is generally not used as a concentrate. However, wheat bran is sometimes added to the diet of a horse for supplemental nutrition, usually moistened and in the form of a bran mash. Wheat bran is high in phosphorus, so must be fed carefully so that it does not cause an imbalance in the Ca:P ratio of a ration. Once touted for a laxative effect, this use of bran is now considered unnecessary, as horses, unlike humans, obtain sufficient fibre in their diets from other sources. Many feed manufacturers combine various grains and add additional vitamin and mineral supplements to create a complete premixed feed that is easy for owners to feed and of predictable nutritional quality. Some of these prepared feeds are manufactured in pelleted form, others retain the grains in their original form.

In many cases molasses as a binder to keep down dust and for increased palatability. Grain mixes with added molasses are usually called "Sweet feed" in the USA and "Coarse mix" in the United Kingdom. Pelleted or *extruded* feeds (sometimes referred to as "nuts" in the UK) may be easier to chew and result in less wasted feed. Horses generally eat pellets as easily as grain. However, pellets are also more expensive, and even "complete" rations do not eliminate the necessity for forage. The average modern horse on good hay or pasture with light work usually does not need supplements; however, horses subjected to stress due to age, intensive athletic work, or reproduction may need additional nutrition. Extra fat and protein are sometimes added to the horse's diet, along with vitamin and mineral supplements.

Soybean meal is a common protein supplement, and averages about 44 per cent crude protein. The protein in soybean meal is high-quality, with the proper ratio of dietary essential amino acids for equids. Cottonseed meal, Linseed meal, and peanut meal are also used, but are not as common. Vegetable oil is a common fat source added to a ration. Corn oil is particularly popular, but other oils are used as well. Rice bran is a very good fat supplement that contains 20 per cent fat as well as fibre and other nutrients. Flax seed is another good source of fat, though it must ground up for horses to digest it. Some commercial feed manufacturers now make products containing both flaxseed and rice bran.

There are hundreds, if not thousands of commercially prepared vitamin and mineral supplements on the market, many tailored to horses with specialized needs. Most horses only need quality forage, water and a salt or mineral block. Grain or other concentrates are often not necessary. But, when grain or other concentrates are fed, quantities must be carefully monitored. To do so, horse feed is measured by weight, not volume. For example, 1 lb. of oats has a different volume than 1 lb. of corn. When continuous access to feed is not possible, it is more consistent with natural feeding behaviour to provide three small feedings per day instead of one or two large ones.

However, even two daily feedings is preferable to only one. To gauge the amount to feed, a weight tape, available at most feed stores, can be used to provide a reasonably accurate estimate of a horse's weight. The tape measures the circumference of the horse's barrel, just behind the withers and elbows, and the tape is calibrated to convert inches or centimeters into approximate pounds or kilograms. Actual amounts fed vary by the size of the horse, the age of the horse, the climate, and the work to which the animal is put. In addition, genetic factors play a role. Some animals are naturally easy keepers, which means that they can thrive on relatively small amounts of food and are prone to obesity and other health problems if overfed. Others are hard keepers, meaning that they are prone to be thin and require considerably more food to maintain a healthy weight.

Veterinarians are usually a good source for recommendations on appropriate types and amounts of feed for a specific horse. There are also numerous books written on the topic. Feed manufacturers usually offer very specific guidelines for how to select and properly feed products from their company, and in the United States, the local office of the Cooperative Extension Service can provide educational materials and expert recommendations. Equids always require forage. When possible, nutritionists recommend it be available at all times, at least when doing so does not overfeed the animal and lead to obesity. It is safe to feed a ration that is 100 per cent forage (along with water and supplemental salt), and any feed ration should be at least 50 per cent forage. Hay with alfalfa or other legumes has more concentrated nutrition and so is fed in smaller amounts than grass hay, though many hays have a mixture of both types of plant.

When beet pulp is fed, a ration of 2 to 5 pounds is usually soaked in water for 3 to 4 hours prior to feeding in order to make it more palatable, and to minimize the risk of choke and other problems. It is usually soaked in a proportion of one part beet pulp to two parts water. Beet pulp is usually fed in addition to hay, but occasionally is a replacement for hay when fed to very old horses who can no longer chew properly. Some pelleted rations are designed to be a "complete" feed that contains both hay and grain, meeting all the horse's nutritional needs. However, even these rations should have some hay or pasture provided, a minimum of a half-pound of forage for every 100 pounds of horse, in order to keep the digestive system functioning properly and to meet the horse's urge to graze.

Recent studies address the level of various non-structural carbohydrates (NSC), such as fructan, in forages. Too high an NSC level causes difficulties for animals prone to laminitis or equine polysaccharide storage myopathy (EPSM). NSC cannot be determined by looking at forage, but hay and pasture grasses can be tested for NSC levels. Concentrates, when fed, are recommended to be provided in quantities no greater than 1 per cent of a horse's body weight per day, and preferably in two or more feedings. If a ration needs to contain a higher per cent of concentrates, such as that of a race horse, bulky grains such as oats should be used as much as possible; a loose mass of feed helps prevent impaction colic. Peptic ulcers are linked to a too-high concentration of grain in the diet, particularly noticed modern racehorses, where some studies show such ulcers affecting up to 90 per cent of all race horses.

In general, the portion of the ration that should be grain or other concentrated feed is 0 to 10 per cent grain for mature idle horses; between 20 to 70 per cent for horses at work, depending on age, intensity of activity and energy requirements. Concentrates should not be fed to horses within one hour before or after a heavy workout. Concentrates also need to be adjusted to level of performance. Not only can excess grain and inadequate exercise lead to behaviour problems, it may also trigger Equine Exertional Rhabdomyolysis, or "tying up," in horses prone to the condition.

Horses normally require free access to all the fresh, clean water they want, and to avoid dehydration, should not be kept from water longer than four hours at any one time. However, water may need to be temporarily limited in quantity when a horse is very hot after a heavy workout. As long as a hot horse continues to work, it can drink its fill at periodic intervals, provided that common sense is used and that an overheated horse is not forced to drink from extremely cold water sources. But when the workout is over, a horse needs to be cooled out and walked for 30—90 minutes before it can be allowed all the water it wants at one time. However, dehydration is also a concern, so some water needs to be offered during the cooling off process. A hot horse will properly rehydrate while cooling off if offered a few swallows of water every three to five minutes while being walked. Sometimes the thirst mechanism does not immediately kick in following a heavy workout, which is another reason to offer periodic refills of water throughout the cooling down period.

Even a slightly dehydrated horse is at higher risk of developing impaction colic. Additionally, dehydration can lead to weight loss because the horse cannot produce adequate amounts of saliva, thus decreasing the amount of feed and dry forage consumed. Thus, it is especially important for horse owners to encourage their horses to drink when there is a risk of dehydration; when horses are losing a great deal of water in hot weather due to strenuous work, or in cold weather due to horses' natural tendency to drink less when in a cold environment. To encourage drinking, owners may add electrolytes to the feed, additives to make the water especially palatable (such as apple juice), or, when it is cold, to warm the water so that it is not at a near-freezing temperature.

SPECIAL FEEDING ISSUES FOR PONIES

Ask a pony owner and they'll tell you, ponies are tougher, smarter, stronger and healthier than any horse. They run on a lot less fuel too. Rations that would starve a horse will keep a pony round and energetic. Most pony breeds developed where the pasture was sparse, the terrain rugged, and the climate harsh. When we pamper them, feed special preparations and lush grass we sometimes do more harm than good. Ponies need only the fraction of the feed that horses do.

Hay for ponies should be good quality grass hay. Your pony probably won't need the nutrition provided by alfalfa and clovers. Lush pasture is a danger zone for ponies. Ponies can founder in less than 60 minutes of grazing if introduced suddenly to lush grass. If you plan to keep your pony on grass introduce it very slowly. Start with 10 minutes of grazing and gradually add a few minutes each day twice a day. If your pony eats too much rich pasture it could lead to such things as colic or founder. You may never be able to leave some ponies on good pasture. Ponies can become obese very quickly and that can lead to health problems.

Good pasture for a pony would be one that he has to work at finding the grass in. Sparse grass that grows slowly would be ideal. Or he could spend a small portion of his time on pasture and the rest in a grassless paddock. Some people use their round pen or a paddock where no grass grows. Another option is to use a grazing muzzle. Make sure your pony has access to clean fresh water. Ponies rarely need concentrates or grain. The exception would be a pony that is working hard: one that is doing several lessons a week, is being driven frequently, is doing something like pulling competitions, or a lactating mare. If your pony is losing condition you could increase the quantity of hay and if that isn't enough add a concentrate that isn't too rich. A forage replacer fortified with vitamins and minerals might give your pony the nutritional boost it needs. Pony mouths are small, so overgrown teeth can be a problem. Don't forget to have your vet check your pony's teeth to ensure it can chew easily. Ponies also need regular de-worming to keep them in the best health.

If you like to feed your ponies even if it is not working hard--and for some owners this is a very satisfying activity--look for a concentrate that is low in calories. Some manufacturers make special pony mixes. These mixes are balanced with the correct amount of supplements for a pony. Don't be tempted to top dress it with a lot of extras like molasses or beet pulp. If you are feeding good hay, the pony is getting a bit of pasture and you have a mineral/salt block available your pony will be getting what he needs. Ponies and miniature horses are usually easy keepers and need less feed than full-sized horses. This is not only because they are smaller, but also, because they evolved under harsher living conditions than horses, they use feed more efficiently. Ponies easily become obese from overfeeding and therefore are at high risk for colic, Equine Metabolic Syndrome, and, especially, laminitis. Fresh grass is a particular danger to ponies; they can develop laminitis in as little as one hour of grazing on lush pasture.

It is important to track the weight of a pony carefully, by use of a weight tape. Forages may be fed based on weight, at a rate of about 1 pound of forage for every 100 pounds. Forage, along with water and a salt and mineral block, is all most ponies require. If a hard-working pony needs concentrates, a ratio of no more than 30 per cent concentrates to 70 per cent forage is recommended. Concentrates designed for horses, with added vitamins and minerals, will often provide insufficient nutrients at the small serving sizes needed for ponies. Therefore, if a pony requires concentrates, feed and supplements designed specially for ponies should be used. In the UK, extruded pellets designed for ponies are sometimes called "pony nuts.".

Special Feeding Issues for Mules and Donkeys

Like ponies, mules and donkeys are also very hardy and generally need less concentrated feed than horses. Mules need less protein than horses and do best on grass hay with a vitamin and mineral supplement. If mules are fed concentrates, they only need about half of what a horse requires. Like horses, mules require fresh, clean water, but are less likely to over-drink when hot.

Fig. Grass hay for Mules and Donkey

Donkeys, like mules, need less protein and more fibre than horses. They do best when allowed to consume small amounts of food over long periods, as is natural for them in an arid climate. They can meet their nutritional needs on 6 to 7 hours of grazing per day on average dryland pasture that is not stressed by drought. If they are worked long hours or do not have access to pasture, they require hay or a similar dried forage, with no more than a 1:4 ratio of legumes to grass. They also require salt and mineral supplements, and access to clean, fresh water. Like ponies and mules, in a lush climate, donkeys are prone to obesity and are at risk of laminitis. Many people like to feed horses special treats such as carrots, sugar cubes, peppermint candies or specially manufactured horse "cookies." Horses do not need treats, and due to the risk of colic or choke, many horse owners do not allow their horses to be given treats. There are also behavioural issues that some horses may develop if given too many treats, particularly a tendency to bite if hand-fed, and for this reason many horse trainers and riding instructors discourage the practice.

However, if treats are allowed, carrots and compressed hay pellets are common, nutritious, and generally not harmful. Apples are also acceptable, though it is best if they are first cut into slices. Horse "cookies" are often specially manufactured out of ordinary grains and some added molasses. They generally will not cause nutritional problems when fed in small quantities. However, many types of human foods are potentially dangerous to a horse and should not be fed. This includes bread products, meat products, candy, and carbonated or alcoholic beverages.

It was once a common practice to give horses a weekly *bran mash* of wheat bran mixed with warm water and other ingredients. It is still done regularly in some places. While a warm, soft meal is a treat many horses enjoy, and was once considered helpful for its laxative effect, it is not nutritionally necessary. An old horse with poor teeth may benefit from food softened in water, a mash may help provide extra hydration, and a warm meal may be comforting in cold weather, but horses have far more fibre in their regular diet than do humans, and so any assistance from bran is unnecessary. There is also a risk that too much wheat bran may provide excessive phosphorus, unbalancing the diet, and a feed of unusual contents fed only once a week could trigger a bout of colic. All hay and concentrated feeds must kept dry and free of mold, rodent feces and other types of contamination that may cause illness in horses. Feed kept outside or otherwise exposed to moisture can develop mold quite quickly. Due to fire hazards, hay is often stored under an open shed or under a tarp, rather than inside a horse barn itself, but should be kept under some kind of cover.

Concentrates take up less storage space, are less of a fire hazard, and are usually kept in a barn or enclosed shed. A secure door or latched gate between the animals and any feed storage area is critical. Horses accidentally getting into stored feed and eating too much at one time is a common but preventable way that horses develop colic or laminitis. It is also important to never give a horse feed that was contaminated by the remains of a dead animal, it is a potential source of botulism. This is not an uncommon situation. For example, mice and birds can get into poorly stored grain and be trapped; hay bales sometimes accidentally contain snakes, mice, or other small animals that were caught in the baling machinery during the harvesting process.

Horses can become anxious or stressed if there are long periods of time between meals. They also do best when they are fed on a regular schedule, they are creatures of habit and easily upset by changes in routine. When horses are in a herd, their behaviour is hierarchical; the higher-ranked animals in the herd eat and drink first. Low-status animals, who eat last, may not get enough food, and if there is little available feed, higher-ranking horses may keep lower-ranking ones from eating at all. Therefore, unless a herd is on pasture that meets the nutritional needs of all individuals, it is important to either feed horses separately, or spread feed out in separate areas to be sure all animals get roughly equal amounts of food to eat. In some situations where horses are kept together, they may still be placed into separate herds, depending on nutritional needs; overweight horses are kept separate from thin horses so that rations may be adjusted accordingly. Horses' teeth continually erupt throughout their life, are worn down as they eat, and can develop uneven wear patterns that can interfere with chewing. For this reason, horses need a dental examination at least once a year, and particular care must be paid to the dental needs of older horses. The process of grinding off uneven wear patterns on a horse's teeth is called *floating* and can be performed by a veterinarian or a specialist in equine dentistry.Diet and Nutrition for Dog There are a great quantity of commercial foods and treats marketed for dogs, and not all are recommended as part of a balanced, healthy diet.

There is some debate as to whether domestic dogs should be classified as omnivores or carnivores, by diet. The classification in the Order Carnivora does not necessarily mean that a dog's diet must be restricted to meat; unlike an obligate carnivore, such as the cat family with its shorter small intestine, a dog is neither dependent on meat-specific protein nor a very high level of protein in order to fulfill its basic dietary requirements. Dogs are able to healthily digest a variety of foods including vegetables and grains, and in fact dogs can consume a large proportion of these in their diet. Wild canines not only eat available plants to obtain essential amino acids, but also obtain nutrients from vegetable matter from the stomach and intestinal contents of their herbivorous prey, which they usually consume.

Domestic dogs can survive healthily on a reasonable and carefully designed vegetarian diet, particularly if eggs and milk products are included. Some sources suggest that a dog fed on a strict vegetarian diet without L-carnitine may develop dilated cardiomyopathy, however, L-carnitine is found in many nuts, seeds, beans, vegetables, fruits and whole grains. In the wild, dogs can survive on a vegetarian diet when animal prey is not available.

Observation of extremely stressful conditions such as the Iditarod Trail Sled Dog Race, and scientific studies of similar conditions has shown that high-protein (approximately 40 per cent) diets including meat help prevent damage to muscle tissue in dogs and some other mammals. This level of protein corresponds to the percentage of protein found in the wild dog's diet when prey is abundant; higher levels of protein seem to confer no added benefit.

Dogs frequently eat grass, which is a harmless activity. Explanations abound, but rationales such as that it neutralizes acid, or that dogs eat grass to induce vomiting to remove unwanted substances from their stomachs, are at best educated guesses. Dogs do vomit more readily than humans, as part of their typical feeding behaviour of gulping down food then regurgitating indigestible material such as bones and fur. This behaviour is typical of pack feeding in the wild, where the most important thing is to get as much of the kill as possible before others consume it all. Individual domestic dogs, however, may be very "picky" eaters, in the absence of this social pressure. Dogs may also appear to eat grass when they are just running the blades through their mouth to gather information. Their sense of smell and taste may act together to detect if other animals have walked through their area or urinated on the grass.

DANGEROUS SUBSTANCES

Human Food

Some foods commonly enjoyed by humans are dangerous to dogs, including chocolate (Theobromine poisoning), onions, grapes and raisins, some types of gum, certain sweeteners and Macadamia nuts. The only known dangerous substance in chocolate is cocoa, so the danger of white chocolate is uncertain. The acute danger from grapes and raisins was discovered around 2000, and has slowly been publicized since then. The cause is not known. Small quantities will induce acute renal failure. Sultanas and currants may also be dangerous. Cooked bones are dangerous for dogs, because the heat of cooking changes their chemical and physical properties so that they cannot be chewed properly. As a result they may splinter into jagged shards that resist digestion. Alcoholic beverages pose comparable hazards to dogs as they do to humans, but due to low body weight and lack of alcohol tolerance they are toxic in much smaller portions.

Household Poisons

Many household cleaners such as ammonia, bleach, disinfectants, drain cleaner, soaps, detergents, and other cleaners, mothballs and matches are dangerous to dogs, as are cosmetics such as deodorants, hair colouring, nail polish and remover, home permanent lotion, and suntan lotion. Dogs find some poisons attractive, such as antifreeze (automotive coolant), slug and snail bait, insect bait, and rodent poisons. Antifreeze is insidious to dogs, either puddled or even partly cleaned residue, because of its sweet taste. A dog may pick up antifreeze on its fur and then lick it off.

Plants

Plants such as caladium, dieffenbachia and philodendron will cause throat irritations that will burn the throat going down as well as coming up. Hops are particularly dangerous and even small quantities can lead to malignant hyperthermia. Amaryllis, daffodil, english ivy, iris, and tulip (especially the bulbs) cause gastric irritation and sometimes central nervous system excitement followed by coma, and, in severe cases, even death. Ingesting foxglove, lily of the valley, larkspur and oleander can be life threatening because the cardiovascular system is affected. Yew is very dangerous because it affects the nervous system. Immediate veterinary treatment is required for dogs that ingest these.

Animal Feces

Dogs occasionally eat their own feces, or the feces of other dogs and other species if available, such as cats, deer, cows, or horses. This is known as coprophagia. Some dogs develop preferences for one type over another. There is no definitive reason known, although boredom, hunger, and nutritional needs have been suggested. Eating cat feces is common, possibly because of the high protein content of cat food. Dogs eating cat feces from a litter box may lead to to Toxoplasmosis. Dogs seem to have different preferences in relation to eating feces. Some are attracted to the stools of deer, cows, or horses.

Other risks

Human medications may be toxic to dogs, for example paracetamol/ acetaminophen (Tylenol). Zinc toxicity, mostly in the form of the ingestion of US pennies minted after 1982, is commonly fatal in dogs where it causes a severe hemolytic anemia.

MAXIMISING DIGESTIBILITY OF FIBROUS FEEDS

SUPPLEMENTATION

A deficiency of any nutrient in the rumen must first affect the rumen ecosystem by reducing microbial growth efficiency which, as the deficiency becomes more acute, will eventually reduce microbial pool size and inevitably digestibility and intake of roughage by ruminants, Decreased digestibility will only be seen with time as the deficiency of the nutrient becomes progressively greater.

Supplementation of a critically deficient animal with nutrients for the rumen will increase microbial growth efficiency, microbial pool size and digestibility of forage which will almost always increase feed intake. The feed intake effect, however, often depends on the degree to which the limiting nutrient is deficient. The most likely deficient nutrients, are non-protein nitrogen, P, S, Mg, Na and also trace elements such as Co, Cu, and Zn. Supplementation with multi nutrients of the type listed above is generally successful in increasing intake and digestibility but the large increases in productivity can be anticipated from combining such supplements with a source of bypass protein.

Sources of Mineral Nutrients and Molasses Urea Blocks

One of the best sources of multiple trace minerals in developing countries is a concentrated plant juice, such as molasses or palm oil sludge. Multi-nutrient blocks are no better than the mixtures of materials from which they are made. Blocks are useful because they form a package readily handled, marketed and used by the farmer with little impact on the time he has for other activities. The major problem with these multi-nutrient blocks occurs in the grasslands where there is a need for appropriate means of encouraging stock to consume them in sufficient amounts.

In these areas molasses is often unavailable or is too expensive. More research is needed to produce suitable packages (*e.g.* blocks without molasses) for such localities.

The importance of block supplementation relates more to the access it allows to livestock managed by small farmers. It may be possible in the future to apply innovations through inclusions in the blocks of chemicals and other materials that will manipulate the rumen (*e.g.* antiprotozoal chemicals) and manipulate the animal (*e.g.* protected chemicals that are absorbed from the intestines) or even control disease and parasitism (anthelminthics). In this respect the production and testing of blocks incorporating anthelminthic is appropriate and being undertaken already by the Commomwealth Scientific and Industrial Organization in co-operation with the Australian Centre for International Agricultural Research (ACIAR).

Improving the Enzymatic Ability of Rumen Microbes to Degrade Fibre

This is an area that has received emphasis and has aroused considerable expectations. The concept is simple, if rumen microbes can be developed with greater abilities to degrade fibre, or its components, the microbes will be able to extract a greater amount of the available nutrients from the feed consumed.

This often receives a high priority because of the concept that animals fed low quality pastures are primarily energy deficient. In fact, these animals more often are inefficiently using the available nutrients. It must be stressed that the application of balanced nutrition is the first priority and if other perturbations of the system are to succeed an animal must have an efficient rumen and also metabolism.

However, in an animal fed a roughage diet, with an efficient rumen and balanced for protein, it is possible that for every 5 units increase in digestibility of the basal roughage resource, there will be 50 per cent increases in liveweight gain on the same basal feed resource.

Silva and Ørskov demonstrated that the extent of breakdown of a fibrous feed was dependent on the rumen medium. Untreated ground straw placed in nylon bags, in the rumen of sheep fed ammoniated straw, was fermented to a greater extent than the same untreated ground straw placed in the rumen of sheep fed untreated straw. This difference might be due to a higher concentration of enzymes in the rumen of sheep fed ammoniated compared to untreated straw.

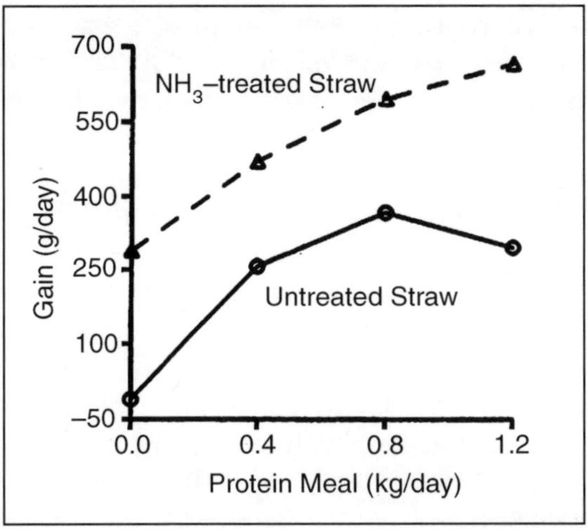

Fig. Growth Response to Different Levels of Protein Meal Supplementation of Cattle given Ammoniated (NH_3-Treated Straw) or Untreated Rice Straw (Untreated Straw)

The two observations discussed above when combined indicate the great potential for improvements in productivity from a successful establishment of a super bug in the rumen, *i.e.* a micro-organism engineered to produce more, or a greater variety, of enzymes for fibre digestion.

Potential Targets for Improving Fibre Digestion

There are potentially three ways to enhance the ability of rumen microbes to degrade fibrous feeds:

1. Selection of microbes, particularly with a high fibrolytic enzyme secretion or that secrete fibrolytic enzymes of high specific activity
2. Creation of microbes with greater spectrum of enzyme secretions by recombinant dna means, *e.g.* Create cellulolytic capacities in xylolytic organisms and vice versa
3. Create microbes with enhanced fibrolytic activity by recombinant DNA means.

In all cases, the prerequisite will be that the organisms are able to grow in and maintain their space in the rumen. These same prerequisites also apply to the production by genetic engineering techniques of high biological value proteins, other secretions of proteins, amino acids and peptides by rumen organisms (see earlier discussion).

State of the Art in Bioengineering of Rumen Organisms

There are number of well established ways of going about incorporating DNA encoding for specific enzymes into anaerobic bacteria by recombinant DNA technologies.

Progress has been relatively swift, with at least 4 groups having now achieved transformation of rumen microbes with foreign DNA. Virtually all transformations have been achieved with plasmid constructs and through electroporation methods. None of the engineered organisms have been reintroduced into the rumen to test their stability and in only one case was the engineered organism a recent isolate from the rumen. The approach of using laboratory strains of rumen organisms which have definitely changed in culture clearly demonstrates that the research is at the basic level in which the transformation was the major object and not the improved digestibility that might result.

A major constraint has been the lack of suitable monitoring systems for identifying the engineered organism within the milieu of the rumen. DNA probes are not proving very appropriate because of the genetic diversity of bacterial strains within a species.

The above discussion is intended to emphasise that their is a great need for considerable basic research before modified organisms that could survive in the rumen is developed. It should also emphasise the long time lag that is likely from the beginning, where virtually no information was available to application.

Selection of Anaerobic Fungi for Better Fibre Degradation in the Rumen

Filamentous structures of the rumen anaerobic fungi penetrate and substantially weaken the xylem cell walls of forages in the rumen, whereas the bacteria do not adhere to, nor degrade these highly resistant plant tissues to any significant extent. Rumen fungi, but not bacteria, penetrate the cuticle barrier that protects the leaf's surface. Research indicates that these fungi have unique enzymes, or enzymes with higher specific activities, that give them the ability to weaken and degrade the most limiting structural barriers to degradation.

On the basis of these observations it appears that research aimed at promoting the biomass of fungi in the rumen or the establishment of highly active rumen fungi (*e.g.* by selection) could be a major step forward in increasing the digestibility of poor quality forages by ruminants. The identification of strains of anaerobic fungi that most rapidly degrade the cell walls of forage plants is a necessary prerequisite to the successful inoculation of fungi into the rumen and which will nutritionally benefit the host.

Culture, selection of strains and the identification of new species of anaerobic fungi, testing of their ability to weaken plant fibre and improve the rate of digestion are worthwhile research areas for development.

The developing countries may be well placed to initiate and develop this research as they have access to animals under extreme dietary conditions that have been adapted over long periods of time and therefore could have already selected fungi for high fibrolytic activity. The research must include developments which allow selected strains to multiply in the rumen and to prevent the competition from wild strains. As rumen fungi are spread by resistant spores which are passed in faeces and then, when taken in by the animal in feed, multiply by sporulation in the rumen, the support of specific species to ensure the survival is a research area with a high priority.

The selection criteria will include the measurement of the rate of solubilisation of fibrous carbohydrates by selected fungi and the weakening of the forage stems following a period of incubation with the forages.

Selection of Bacteria for Fibrolytic Activity

Selection of bacterial strains for improved cellulolytic activity appears to be less vital as it is the fungi that are relatively more important in the breakdown of the most resistant components of the forages, particularly the low quality forages.

Solubilisation of Lignin

Lignin is found within the cell walls of forage plants where it is closely associated with hemicellulose, forming a matrix surrounding the orderly cellulose microfibrils. Lignin is also found in high concentrations in the middle lamella between fibres where it functions as a binder for contiguous cells. Polymeric lignins are not degraded by any known anaerobic organisms. The aerobic white rot fungi *Basidiomycetes* degrade lignin more rapidly and extensively than any other studied microbial group. It is generally accepted that the delignification of a ligno-cellulose material will increase the rate of attack upon the cellulose/hemicellulose compounds when this enters the rumen.

Under aerobic conditions there are two mechanisms of biological delignification:

1. Mineralisation, with lignin being converted directly to carbon dioxide
2. Solubilisation when lignin is released (solubilised) from ligno-cellulose with a variable amount of hemicellulose material attached.

Solubilisation of lignins of grasses is particularly important for its degradation in soils. Something similar to solubilisation seems to occur in the rumen, as considerable lignin appears to be precipitated in the abomasum and therefore, is soluble in the rumen fluid (this is often referred to as acid-precipitatable polymeric lignin (APPL)). The enzymes for solubilisation of lignin, if present in rumen organisms, may be capable of being boosted by genetic means or by selection. In the event that they are not present in rumen organisms, the potential for inducing lignin-solubilising enzymes becomes a major possibility for improving digestibility of fibre in the anaerobic ecosystem of the rumen.

Detoxification of Anti-microbial, Toxic and Antiquality Factors in Plants

Plants in order to survive insect, fungal and bacterial attack, have developed secondary compounds which detract from these organisms colonising the leaf tissues. Some shrubs and trees respond to leaf damage as occurs by grazing and produce greater quantities of secondary compounds which often make them inedible. Recently it has been shown that the grazing of an Acacia tree caused it to release ethylene which within minutes, caused the adjacent acacias in the group to commence synthesising tannins which then caused the grazing deer to cease consumption of the leaves of these trees.

Many secondary plant compounds are toxic to the rumen microbes or directly or indirectly toxic to the animal following microbial metabolism in the rumen. This restricts the feed base of ruminants and, at times, prevents stocking of land. Goats and probably other grazing ruminants are able to detoxify many of these secondary plant compounds in the liver.

FACTORS INFLUENCING EFFICIENCY OF FEED UTILISATION

Digested feed clearly has a major effect on nutrient supply, the most important factor governing the extent to which cows can meet their energy, protein and other nutrient needs is the amount of feed they consume. This is especially important in the early stages of lactation when the energy demand for production is higher than intake can support, creating an increasingly negative energy balance which cows have to meet from body reserves, milking off their backs.

The efficiency with which absorbed nutrients are converted to ruminant products (liveweight, milk etc.) is dependent on precisely meeting the animal's requirements above maintenance for individual nutrients required for the particular function. These, at times, are influenced by body condition as affected by previous health and nutritional history, the demands for body temperature control and the requirements for substrate oxidation in exercise (or work).

Graham *et al.* showed that the quantitative oxidation of individual nutrients (largely fat) depended on the degree of heat/cold stress of the animal. A cetogenic substrate was largely used to keep an animal warm when it was required to increase its metabolic rate in cold stress. In this report it is assumed that a cold stressed animal will oxidise acetogenic substrate for heat production until 'surplus' acetogenic substrate is totally utilised, after which fat mobilisation provides an extra and often the major source of metabolic fuel.

The apparently preferential oxidation of circulating acetate leaves a higher ratio of amino acids (and glucose) in the nutrients available for production than would be available to an animal in its zone of thermoneutrality. Conversely, an animal that is not cold has more acetogenic substrates available for anabolic purposes. The environment can, thus, alter the partitioning of nutrients into productive functions and therefore affect the efficiencies of feed utilisation. The design of supplements, to balance diets for ruminants, needs to account for the varying demands for nutrients brought about by the thermal environment of the animal.

It is recognised that cold stress in animals often increases voluntary feed intake and rumen turnover rate. And in this way on some diets it increases microbial cells moving to the lower tract and this has been suggested to increase the P/E ratio in the nutrients available for maintenance or production.

As an example of how environmental factors can change the nutrient balance available to ruminants for anabolism and maintenance, a model used previously to predict the relative availability of specific nutrients to a "standard steer" has been modified to use with sheep and includes the effects of cold stress. The model is based on the sheep (closely shorn) used in the studies of Graham *et al.* which were fed on a daily basis 600, 1200, or 1800 g of a dried grass pellets and subjected to short periods at environmental temperatures ranging from 8 to 40°C at a relative humidity of 50 per cent.

The data in Table indicate that the need to maintain body temperature may require a considerable proportion of the available acetogenic nutrients to be oxidised. In the absence of a cold stress considerably more of digestible nutrients (and in particular more acetogenic substrate) is available for maintenance and synthesis. If in thermoneutral conditions the acetogenic substrate cannot be utilised for synthesis of tissue component because of a low availability of essential amino acids and/or glucose (*i.e.* imbalance in P/E or G/E) then acetate must be dissipated as heat. If the animal is able to oxidise the excess substrate but cannot dissipate the heat-generated because environmental temperature and humidity are high then it could allow its body temperature to increase to some extent but eventually it must reduce its feed intake. If the animal's body temperature rises, metabolic rate increases through the oxidation of protein which may have implications for protein requirements of ruminants in the tropics and differential responses to supplementation in the tropics as compared to temperate areas.

The temperature/humidity at which ruminants are cold stressed depends greatly on the level of feed intake, the insulation provided by the hair or wool coat and the environmental conditions prevailing, *e.g.* wind, rain and availability of shelter. Thus, the environmental temperatures at which minimum extra heat production to combat the cold stress occurs will probably move through a range of from around 10°C to 40°C.

CLIMATE, SUPPLEMENTATION AND INTAKE OF "LOW QUALITY FORAGES"

There has been vigourous debate on whether supplementation of sheep and cattle on low quality forage based diets with urea and/or bypass protein increases intake of the basal feed resource. The differences in results may be hypothesised to be a result of an interaction between climate and the balance of nutrients available from a diet. When research results (Australian) on the effects of supplementation to balance nutrition of cattle on low quality forages are grouped according to climatic zones a pattern emerges.

It appears to be in the tropics and subtropics where poor quality forage intake by cattle is low without supplementation and where significant responses in feed intake occurs when a non-protein nitrogen deficiency is corrected and extra protein that escapes rumen fermentation is provided in the diet.

It is strongly stressed that supplementation with urea and proteinmeals increases voluntary intake of poor quality forages by cattle under tropical conditions to approximately the same level of intake as unsupplemented cattle under temperate conditions. In this situation the supplement is only correcting a depressed intake back to normal intake.

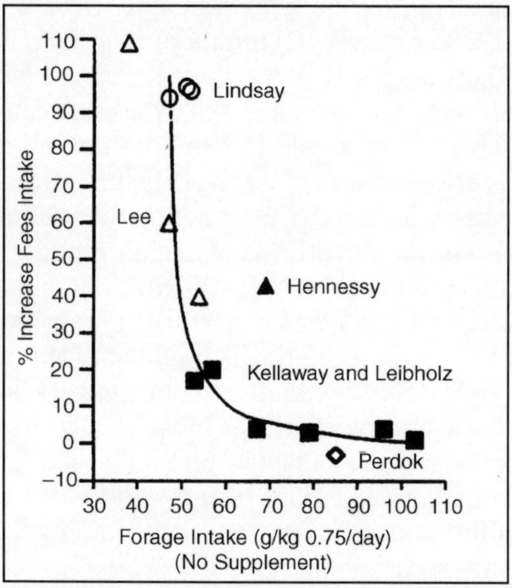

Fig. Intake of Low Digestibility Forages by Cattle Either Unsupplemented or Supplemented with Bypass Protein or Bypass Protein and Urea

The conclusion that can be drawn from this is that supplements which improve the P/E ratio in nutrients absorbed by cattle fed low quality forage reduces metabolic heat production.

Where metabolic heat production in unsupplemented cattle fed low quality forages would increase body temperature then the animal reduces its feed intake. This reduction in voluntary feed intake is ameliorated by the supplement which allows the acetogenic substrate which would otherwise have to be oxidised to be partitioned into synthetic reactions with a resultant decrease in heat production. The concept of small increases in P/E ratio being able to reduce metabolic heat and at times therefore allow an animal to consume more food might explain the effects of increasing levels of urea in a diet on forage intake (when digestibility is no longer increased) and also the occasional effects on feed intake of branched chain VFA supplements.

The concept is that it is a supplement that improves microbial growth efficiency which has an effect on feed intake and this is only seen in the hot conditions when feed intake is depressed.

The interaction of nutrition and climate may explain why there is a stubborn disbelief by some researchers from developed countries (largely in the temperate areas) of research carried out in developing countries in the tropics. Many of the results of supplementation indicate that a protein that escapes rumen fermentation stimulates both the level and efficiency of production of milk (or live-weight gain) in ruminants fed on crop residues.

The discussion above indicates that ruminants in hot countries have an advantage of not having to oxidise much acetogenic substrate (or body fat) to keep warm. By balancing the diet with supplements, this acetogenic substrate may be captured in products or oxidised to provide ATP for assimilation of the additional nutrients into products. In cold/cool countries supplementation with protein is less necessary, as the utilisation of surplus acetate for heat, decreases the need to balance nutrients. As long as feed intake is high (*i.e.* the diet is highly digestible and perhaps cold stress stimulates intake) production remains relatively high as the nutrients for heat production are extracted and the balance used in synthetic reactions.

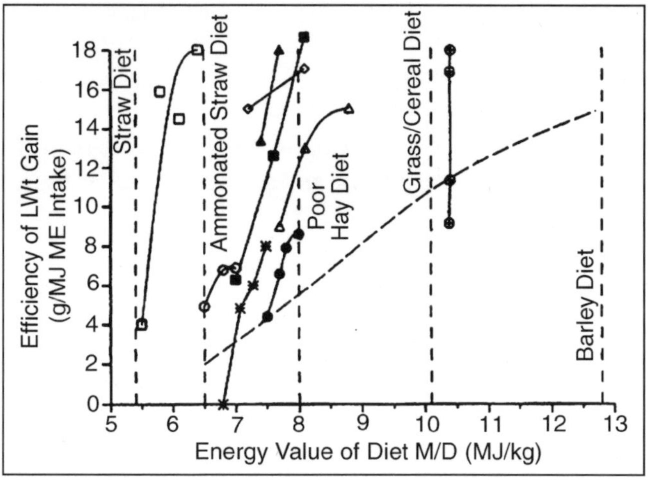

Fig. Schematic Relationship between Diet Quality (Metabolisable Energy mj/kg Dry Matter) and Food Conversion Efficiency (G Liveweight Gain/mj me) (---). The Relationships Found in Practice with Cattle Fed on Straw or Ammoniated Straw with Increasing Level of Supplementation. Australia (%, o, •), Thailand (") and Bangladesh (%). Recent Relationships Developed for Cattle Fed Silages Supplemented with Fish Proteins (o) and Tropical Pastures Supplemented with Cottonseed Meal (*) are also Shown. This Illustrates the Marked Differences that Result when Supplements High in Protein are given to Cattle on Diets of Low me/kg dm

Nevertheless, increases in the efficiency of utilisation are obtained when low protein diets are supplemented with a bypass protein even in temperate countries and at times feed intake is also stimulated but it is unknown whether the animals in such studies were actually subjected to hot conditions.

It can be concluded that ruminants in the tropics that are adequately supplemented with small quantities of essential nutrients may produce at the same rate on a lower digestibility feed as an animal on higher digestibility feed in a cold environment.

To emphasise the differences in potential thermal stress of animals under different conditions the average temperature humidity index (THI) (which is an index of potential heat stress conditions for ruminants on a monthly basis for Cambridge (England), Chittagong (Bangladesh), Bangkok (Thailand) and Armidale (Australia) are shown in Figure.

The critical THI (72) for high milk producing cows as determined by Johnson (1987) is included in the figure. However, it must be emphasised that in addition to temperature/humidities,. the critical THI will depend on the insulation provided by the animal's coat and its behaviour in seeking shelter, as well as the incidence of wind and rain in addition to level and quality of feed intake.

Many studies have shown that at the same forage intake by ruminants with an already efficient digestion that a supplement of protein that reaches the small intestine increases the efficiency of feed or metabolisable energy utilisation for growth.

This is positive proof that wasteful oxidation of nutrients can occurs. It seems reasonable that, because Blaxter and his colleagues showed that acetogenic substrate are largely "burned off" that the inefficiency of ruminants on forage based diets is a result of acetate being oxidised wastefully. This points to a major difference in considering the nutrition of ruminants in the tropics as compared with temperate countries.

ANIMAL PRODUCTION BUILDINGS

EMISSION MEASUREMENT

Emission refers to the rate at which gases or particulates are being released into ambient air. It is also a mass flux per unit area and time from a particular surface. This is in contrast to concentration-only measurements. Emission rates are determined by multiplying the concentration of a component by the volumetric flow rate at which a component at a given concentration is being emitted.

Surprisingly, while accurately measuring gas and odour concentrations within facilities is feasible, the determination of building or manure management system emissions is not straightforward. For example, it is not sufficient to count the number of fans, multiply by some average fan ventilation rate, and then multiply by the gas concentration.

Likewise, it is not sufficient to estimate mass flux of a specific gas from the surface of litter on a floor, or manure within the facility, and then assume the building emission is constant regardless of the number of fans running; nor would it be appropriate to assume all similar facilities exhibit similar emissions. While these aforementioned *crude estimates* might be suitable for a rough "ball-park" estimate of building emission, at best they would be only useful for that point in time and they completely neglect the effect of daily husbandry activities (feeding, lights, etc) and disturbances to the thermal control systems (especially weather systems). Odour and gas emission rates are often normalized to the number and weight of animals by dividing the total emission rate by the number of animal units (AU), where one AU is equal to 500 kg of animal live weight.

Emission expressed in terms of AU is often referred to as the emission factor. Area-specific emission, or flux rate, is determined by dividing the total emission rate by the emitting surface area. Thus the comparison of emissions from various studies is often difficult if not done on the same basis, such as AU, animal live weight, animal place, area, or volume or weight of manure. Furthermore, the definitions of AU and animal place are not standardized. Therefore, conversion of emissions reported in one study to the units used in another study is not always possible; and when done, may lead to misleading interpretations.

Also, data collection periods vary widely, ranging from a few hours to several days. In some cases units from original data sources were converted to grams of compound per AU and per day for comparison purposes, but this may not fully correspond to actual emission measurements. Conversion of daily to annual emission values is not encouraged as emission rates vary widely during the year depending on season, air temperature, humidity, etc.

Ventilation Rates

A major impediment to determining emissions is the difficulty in knowing how much air is being exchanged. Mechanically ventilated facilities typically use a large number of fans and if the interior airspace is not well-mixed then gas concentration and hence emission rate may differ at each fan. Accurate measurement of airflow is difficult, and a number of factors commonly found in poultry and livestock facilities make this especially so, including dust accumulation on shutters and blades, loose belts, loss of building static pressure which results in variable ventilation effectiveness, and poor mixing, etc. Basically, three methods can be used for determining building ventilation rates. One method, used for *in situ* ventilation measurement, has been developed by Simmons *et al*. and has been used in poultry facilities. The device is a motorized anemometer array controlled and monitored with a computer.

It uses five propeller-driven DC generators mounted on a horizontal bar or rack. The bar travels vertically and the instruments perform an equal area traverse. Volumetric flow determinations can be made in either vertical direction (*i.e.* going up or down). Following the traverse, the total fan output is calculated as a function of the area of the opening of the anemometer array. Its accuracy has been shown to be within 1 per cent when used with 122 cm diameter fans. The second method uses heat production data and its relation to animal carbon dioxide (CO_2) production. This latter quantity is measured and the building ventilation rate is obtained by inverse solution of a building CO_2 balance. In addition to these two techniques, measurement of the building's static pressure may be used if fan manufacturer's performance data are available and if the fans are in a condition similar to the standard test fans used in the performance tests.

European studies on gas emissions from livestock and poultry facilities, often estimate building ventilation rates derived from the relationship between metabolic heat production and the CO_2 production of the animals and manure (if stored in a deep pit, underneath the animals). The validity of this method is based on two factors: a) valid heat production values for the animals, and b) CO_2 production is solely from respiration of the animals. The use of certain literature heat production data, mostly dating back 20 to 50 years, has been questioned because of the drastic advancement in animal genetics and nutrition.

Moreover, depending upon the manure handling systems, the measured CO_2 production can contain considerable contribution by microbial activities of the manure (*e.g.*, manure storage in a high-rise building or deep-pit system). Therefore, building ventilation rates derived with the latest heat production data from intensive laboratory measurements should be more reflective of the modern genetics, nutrition, and manure management practices. Although this technique is less accurate than ventilation flow rate measurement, it has the advantage of being applicable in principle to both mechanically and naturally ventilated buildings.

Odour

Odour emissions from animal production sites are one of the most important factors to consider when determining setback distances from neighbours since the human nose can readily detect odours. Furthermore, odours are often perceived as indicators of airborne pollutants. Livestock and poultry odours originate from four primary sources: animal buildings, feedlot surfaces, manure storage units, and land application of manure. Of these four sources, land application of manure is probably the biggest source of odour emissions and complaints.

Although not typical, daily land application of manure is still practiced by some producers. Irrigation of manure is still also practiced throughout the

United States, in spite of the significant emissions of odour and gases this practice generates. It should be noted that irrigation of anaerobic lagoon liquid generates fewer odours than irrigation of liquid manure, but odour intensity can be high when liquid from heavily loaded lagoons is irrigated as compared to lightly loaded lagoons. Unfortunately, very little scientific information is available on odour emission from manure irrigation. In the Midwest, particularly in the corn belt area, land application typically occurs during specific periods of the year (usually in the fall, but spring application is also practiced) and known odour control management practices, such as injection of liquid manure into the soil, are available to minimize odour emissions. Therefore, emissions from land application are concentrated in short periods of time and may not be such a nuisance as compared to continuous and long duration emissions from other sources such as animal housing, feedlot surfaces, manure storage, and treatment units.

This may help partially explain the fact that odour emission rate measurements have been and continue to be primarily measured from animal housing facilities and manure storage units. Most livestock and poultry odours are generated by the anaerobic decomposition of livestock wastes such as manure (feces and urine), spilled feed, bedding materials, and wash water. The organic matter in these wastes is microbially transformed into non-odourous end products under aerobic conditions. However, in anaerobic environments, the decomposition of organic compounds results in the production of odourous volatile compounds that are metabolic intermediates or end products of microbial processes. Many of these compounds are then carried by ventilation air, airborne dust, and other particles and dispersed into the atmosphere.

Odour must first be quantified to determine odour emission values. Air samples are diluted with a known amount of odour-free air. The dilutions are presented to a specially trained panel of test personnel using an olfactometer, which is an air dilution device. The odour detection threshold (ODT) is the number of dilutions with odour-free air required for an odour to be perceived by 50 per cent of the panel members.

One odour unit (OU) is defined as the amount of odourant at the panel ODT and is dimensionless. However, the ODT of a sample is often expressed as odour units per cubic meter (OU m^{-3}) for calculation convenience of odour emission. If this convention is followed, then odour emission rates (OU s^{-1}) from a livestock building or manure storage unit are the product of the ventilation airflow rate (m^3 s^{-1}) through the barn or over the storage and the odour concentration (OU m^{-3}) in the exhaust air. Few researchers have attempted to quantify odour and gas emission rates from animal housing and results are widely variable.

Table lists odour flux rates measured from buildings for various animal species. This variation likely stems from the lack of standardized methods used to measure both odour and emissions. For example, air samples are often collected and stored in TedlarÔ bags until evaluation by dynamic olfactometry can be performed. However, Zhang *et al.* reported that these bags emitted significant levels of acetic acid and phenol, which are common odourants found in livestock and poultry manure. In addition, the TedlarÔ bag was found to have an absorptive selectivity for certain odourants such as indole and skatole. The white paper on odour mitigation for concentrated animal feeding operations gives a detailed description and discussion of odour sampling and measurement. Lim *et al*. evaluated odour emission and characteristics at two commercial swine nurseries during the spring. Five sampling visits were made to each nursery and nine or ten air samples were collected during each visit.

Zhu *et al*. measured odour at seven different facilities to determine daily variations. Air samples were collected every two hours over a 12-hour period during the day. Watts *et al.* measured odour emissions from a feedlot pen using a portable wind tunnel over a five-day period following 64 mm of rain. The highest emission occurred about 48 hours after the last rainfall. The peak odour concentration was about 60 times higher than odours from the dry pen.

Table. Odour Flux Rates from Animal Housing

Species	Production unit	Location	Odor Flux Rate OU m^2 s^{-1}	Reference
Pigs	Nursery (deep pit)	Indiana	1.1-2.7	Lim et al. (2002)
	Nursery	Minnesota	7.3-47.7	Zhu et al. (2000b)
	Finishing	Minnesota	3.4-11.9	Zhu et al. (2000b)
	Farrowing	Minnesota	3.2-7.9	Zhu et al. (2000b)
	Gestation	Minnesota	4.8-21.3	Zhu et al. (2000b)
	All types	Minnesota	0.25-12.6	Gay et al. (2002)
Poultry	Broiler	Minnesota	0.1-0.3	Zhu et al. (2000b)
	All types	Minnesota	0.3-3.5	Gay et al. (2002)
Dairy	Free-stall	Minnesota	0.3-1.8	Zhu et al. (2000b)
	All types	Minnesota	1.3-3.0	Gay et al. (2002)
Beef	Feedlot	Minnesota	4.4-16.5	Gay et al. (2002)
	Feedlot	Australia	12.5-725	Watts et al. (1994)

Ammonia

Ammonia is colorless, lighter than air, highly water-soluble, and has a sharp, pungent odour with detection threshold between 5 and 18 ppm. Gaseous NH_3 has a mean life of about 14 – 36 hours depending on weather. NH_3 is classified as a particulate precursor, *i.e.* in the vapor phase it will react with other compounds to form particulates. NH_3 and chemical combinations (NH_x) are important components responsible for acidification in addition to sulfur compounds (SO_x), nitrogen oxides, and volatile organic components. Ammonia is deposited downwind of sources by both "dry" and "wet" methods, with dry deposition generally occurring locally. In fact, the amount of ammonia deposited locally is shown to be quite dependent on downwind land-cover with transport and deposition being quite variable across the landscape. Other research has shown that local deposition is concentrated in the first 500 meters from the source.

Ammonia may cause several ecological problems in the environment. First, excess inputs of nitrogen may lead to considerable changes in plant communities with the result that plants which prefer low nitrogen soils disappear and there is an increase in nitrogen indicator plants. Second, acidification in soils with low buffer capacity may occur after nitrification of the added nitrogen.

A falling pH leads to the dissolution of toxic soil constituents such as aluminum ions, and to the leaching of nutrients and aluminum into the groundwater. Third, the natural capability of forest soil to take up methane (CH_4) is decreased by NH_3 deposition, thus increasing the concentration of CH_4 in the atmosphere. Fourth, surface waters may be affected by eutrophication and acidification.

Finally, NH_3 depositions on buildings will promote bacterial growth, which contributes substantially to weathering and corrosion damage of the buildings. The white paper on ammonia emissions from animal feeding operations gives a more detailed description of the environmental impacts of ammonia from animal production.

Ammonia release from animal sources is prevalent due to the inefficient conversion of feed nitrogen to animal product. Livestock and poultry are often fed surplus nitrogen with high protein feeds to ensure nutritional requirements are met. Nitrogen that is not metabolized into animal protein is excreted in the urine of swine and cattle and in the uric acid excreted by poultry. Further microbial action releases NH_3 to the atmosphere.

Ammonia levels of 5 to 10 ppm are typical in well-ventilated swine confinement buildings where slatted floors allow manure to fall into underground manure storage pits. Concentrations of NH_3 tend to be slightly higher (10 to 20 ppm) in buildings where manure is deposited on solid floors. NH_3 levels in animal housing can exceed 25 ppm when lower winter ventilation rates are used and can reach 40 ppm in poorly ventilated buildings or in the manure storage area of high rise layer houses.

Very high levels of NH_3 concentrations, such as 2,500 ppm may be fatal. The U.S. Occupational Safety and Health Administration (OSHA) indoor 8-h NH_3 exposure threshold is 25 ppm, which is similar to NH_3 threshold limits in many other countries.

Table. Ammonia Emission Factors from Livestock and Poultry Housing

Species	Production unit	Notes	Emission Factor g NH_3 AU^{-1} day^{-1}	Reference
Pig	Finish	Partly slatted	42	Aarnink et al. (1995)
	Finish	Litter	34-90	Groot Koerkamp et al. (1998a)
	Finish	Litter	50-62	Groot Koerkamp et al. (1998a)
	Finish	Fully slatted	72	Hinz and Linke, (1998)
	Finish	Fully slatted	128	Demmers et al. (1999)
	Finish	Fully slatted – no pigs	5-8	Ni et al. (2000)
	Finish	Slurry removed weekly	30	Osada et al. (1998)
	Finish	Fully-slatted	32	Osada et al. (1998)
	Finish	Fully slatted	40-50	Ni et al. (2000)
	Finish	Fully slatted (warm weather)	68-274	Ni et al. (2000)
	Finish	Fully slatted	10-80	Zhu et al. (2000a)
	Finish	Fully slatted	310	Zahn et al. (2001)
	Gestation	Litter	18-78	Groot Koerkamp et al. (1998a)
	Gestation	Slats	25-40	Groot Koerkamp et al. (1998a)
	Gestation	Fully slatted	2.2	Zhu et al. (2000a)
	Nursery	Slats	15.6-37.4	Groot Koerkamp et al. (1998a)
	Nursery	Fully slatted	23-160	Zhu et al. (2000a)
Poultry	Layer	Winter	190	Wathes et al. (1997)
	Layer	Summer	300	Wathes et al. (1997)
	Layer	Deep litter	177-261	Groot Koerkamp et al. (1998a)
	Layer	Battery	14-224	Groot Koerkamp et al. (1998a)
	Broiler	Winter and Summer	216	Wathes et al. (1997)
	Broiler	Litter	53-200	Groot Koerkamp et al. (1998a)
	Broiler	Litter	45	Demmers et al. (1999)
	Broiler	Litter	5.8-8.4	Zhu et al. (2000a)
Beef		Straw bedding	8.9-21.6	Groot Koerkamp et al. (1998a)
		Slats	8.4-16.6	Groot Koerkamp et al. (1998a)
		Straw bedding	19.4	Demmers et al. (1998)
		Feedlot	18.3-67.7	Hutchinson et al. (1982)
Dairy		Straw bedding	6.2-21.4	Groot Koerkamp et al. (1998a)
		Free stall	20.2-42.5	Groot Koerkamp et al. (1998a)
		Free stall with straw	31.7	Demmers et al. (1998)

A recent ammonia emission inventory from UK agriculture estimated emission as 197 kt NH_3-N year^{-1}. Emissions from livestock and poultry housing accounted for 7 per cent, 12 per cent, and 19 per cent for pigs, poultry, and cattle, respectively. Table lists published ammonia emissions from livestock and poultry housing.

Ammonia emissions from beef feedlots and dairy facilities appear to be less variable and lower than NH_3 emissions from swine and poultry housing. However, the limited number of data from beef and dairy operations may account for the low range in values.

Currently, there is wide disparity between the few published tabulations of both swine and poultry emission factors. Ammonia emission factors from swine housing units vary from 0.09 to 12.9 g NH_3 AU^{-1} hr^{-1}, where AU is an animal unit corresponding to 500 kg body mass. Numbers from pig finishing units appear to be higher than both gestation and nursery facilities. Measurements from poultry facilities indicate that ammonia emission factors vary 50-fold, from 0.24 to 12.5 g NH_3 AU^{-1} hr^{-1}. Emission factors from layer facilities seem to be consistently higher than those from broiler facilities.

A recently completed U.S. EPA funded study, stated that livestock (including poultry) contribute 50-70 per cent of the total national ammonia emission inventory, which is about 5,300 kt/year. However, the underlying emission factors for different livestock and poultry types were taken from a systems analysis with limited U.S. agricultural input and yet were used to extrapolate to an entire national level.

For example, the US-EPA estimated annual emission for layer hens is approximately 435g NH_3 per bird (which can be traced back to the Battye report). By contrast, The Netherlands currently use a range of 10-83 g NH_3 per bird annual emissions. Considering that there were on average 322 million layers in the U.S. in 1999, the difference between the 83 and 435 g estimates results in a disparity in annual contribution to the national annual inventory of roughly 113,300 metric tons of NH_3. This example clearly indicates that the lack of quality, scientific-based emission data may result in system models that predict highly inaccurate estimates of NH_3 emission contribution by animal production.

The limited number of NH_3 emission data for beef and dairy facilities show a narrower range and significantly lower values as compared to swine and poultry. Gay *et al.* have recently summarized NH_3 flux rates from 66 farms in Minnesota. Swine housing means varied from 0.35 to 13.0 g NH_3 m^{-2} day^{-1}, poultry housing from 2.85 to 8.0 g NH_3 m^{-2} day^{-1}, dairy about 3.7 g NH_3 m^{-2} day^{-1}, and beef feedlots from 2.2 to 4.4 g NH_3 m^{-2} day^{-1}. Ventilation rates from mechanically ventilated buildings were determined by measuring static pressure across the fan. For naturally ventilated buildings a CO_2 mass balance approach was used. It is difficult to compare this data to other studies because it is highly variable and not reported on the basis of animal units. However, the data indicate that NH_3 emissions from swine and poultry housing are consistently higher than NH_3 emissions from dairy and beef housing and open feedlots.

FOOD REQUIREMENTS

VITAMINS

The fact that human beings were subject to various diseases as a result of deficient diet has been long known. Even the early Egyptians knew that night blindness could be cured by eating the livers of domestic animals. Scurvy occurred among the crusaders of the thirteenth century. In 1520 the Austrian surgeon, Kramer, pointed out that scurvy could be cured by oranges, limes, or lemons, and in 1795 lemon juice was introduced in the English navy as a preventive of the disease.

In 1882 the Japanese admiral Takaki (or Takagi) was able to cause a marked decrease in the number of cases of beriberi in the Japanese navy by improving the diet of the sailors. Vitamin research really began with the discovery that animals could be used in experimentation with deficiency diseases.

The first clear-cut experimental work is that of Eijkman. He found that if chickens were fed polished rice they developed a disease, called polyneuritis, very similar to beriberi. This disease could then be cured by feeding the birds the polishings which had been removed from the rice. Following this important experiment, progress was for a time slow, and most workers in the field of nutrition did not realise its significance.

Hopkins was one of the exceptions. As far back as 1906 he noted that no animal could live on a diet of pure protein, fat, and carbohydrate, and that obscure dietary factors were responsible for such diseases as scurvy and rickets. It was not until 1915 that it was realised that more than one type of vitamin was involved in nutrition.

In that year McCollum and Davis showed that two types of unknown substances were necessary for the normal nutrition of the rat. At first these were called "fat-soluble A" and "water-soluble B," but it was not long before the term vitamin was accepted throughout the world, and fatsoluble A and water-soluble B became known as vitamin A and vitamin B. Soon there came to be a growing alphabet of vitamins. Then followed a period in which vitamin B began to be subdivided into B1, B2, etc.

At the present time it is customary to speak of the vitamins A or the vitamins of the B complex. As we come to know the true chemical composition of the individual vitamins, there is an increasing tendency to call them by their correct chemical names instead of by letters. Thus vitamin C is now commonly referred to as ascorbic acid.

There is no clear-cut definition of vitamins, no sure way of deciding whether a substance important in nutrition is to be regarded as a vitamin or not. Vitamins are organic substances capable in small concentration of affecting the health or influencing the rate of growth of organisms. There are numerous substances that act in this way. Some of them are regarded as vitamins, others not, or there may be differences of opinion.

Thus it was noted in the last chapter that certain unsaturated fats were necessary for the growth and well-being of rats. These fats can be regarded as a fat requirement or they can be called vitamin F. Similarly, various insects require cholesterol.

This may be regarded simply as a sterol requirement, or the cholesterol may be considered a vitamin in the same sense that calciferol (vitamin D) is a vitamin. The fact that vitamins are capable of exerting a powerful influence when present in extremely low concentrations is an indication that they act in some way as catalysts.

Certainly this is true of some vitamins. It is by no means an easy task to present an up-to-date summary of our information concerning vitamins.

New papers are being published at a rapid rate, and it is well-nigh impossible for anyone not a specialist in the field to keep abreast of them.

For the student, perhaps the easiest way to become informed is to read some book on the subject or some recent treatise on physiological chemistry. The earlier studies of vitamins were essentially practical. Men suffered from diseases due to vitamin deficiency. It was important to discover which foods contained the necessary vitamins, and in what amount. Lacking a chemical knowledge of the vitamins it was essential to develop biological methods of assay.

Thus, it was found that a certain amount of vitamin was necessary to cure pigeons of polyneuritis or to permit normal growth in rats. On the basis of such studies on rats, pigeons, and guinea-pigs, various types of vitamin units were established, and the vitamin content of all sorts of foods was determined.

To raise colonies of rats and guinea pigs is rather an expensive and time-consuming business. Fortunately, the presence of vitamins and the amount of a certain vitamin contained in a given sample of food can be tested with organisms much easier and less expensive to raise than mammals and birds. Insects require many of the known vitamins and they might be used as test animals. But much more suitable are various bacteria and fungi. They can be raised in test tubes in a minimum of space. Obviously one must first determine the exact vitamin requirements of a given type of bacterium or fungus.

The assay of vitamins by the use of bacteria is now well established practice and there is a large literature in the field. This has been reviewed by Snell. In 1941, Beadle and Tatum by x-ray treatment were able to produce mutant strains of the fungus Neurospora which are highly useful for vitamin assay. The usual type of Neurospora requires only the vitamin biotin, but by the use of x-rays and ultraviolet radiation within a few years mutants were obtained which had a specific requirement for every one of the B vitamins then known with the single exception of folic acid. Similar mutants have also been obtained in bacteria. The chemical composition of most vitamins is now known. This makes possible direct chemical analysis.

Often such analysis is aided by spectroscopic study. Thus the A vitamins give a blue colour with antimony trichloride and this colour is the basis for spectroscopic tests. Some vitamins give characteristic absorption bands in the ultraviolet. Likewise, some vitamins are fluorescent, and the fluorescent colour can be tested spectroscopically.

Thus in addition to ordinary chemical assay methods, spectroscopic tests are of great value. The subject is discussed at length in volume 1 of György *Vitamin Methods*. In many cases a vitamin may be replaced by other chemical substances related to it chemically. Thus various substances related to ascorbic acid (vitamin C) behave like ascorbic acid in preventing scurvy.

There are many substances with vitamin D or vitamin K activity. The fact that a substance has a chemical structure similar to that of a vitamin does not necessarily indicate that the substance will act like the vitamin. Often, indeed, a substance which resembles a particular vitamin in its chemical composition is an antagonist to it and tends to prevent its action. So, for example, Dicumarol is chemically close to vitamin K; and yet whereas feeding of vitamin K favours blood clotting in higher animals, Dicumarol prevents clotting. Similarly, various analogues of pyridoxine (a B_6 vitamin) act to prevent its action. Other cases of anti-vitamins could be cited.

Studies of antivitamin action may give clues as to the mechanism of the action of the vitamin they antagonize. The subject is also of considerable interest to the theoretical pharmacologist. More and more cases are being discovered in which the harmful action of a particular drug is due to its chemical affinity to some substance important for the vital process. In some cases, the need for a vitamin is lessened by feeding a substance which can be converted into it. In some organisms, tryptophan can be converted into nicotinic acid. Thus, chicks do not require nicotinic acid if fed enough tryptophan. Moreover, isotope studies show that tryptophan can be converted into nicotinic acid.

Another instance in which a necessary vitamin can be replaced by an amino acid has been described for the lactic acid bacteria, Streptococcus faecalis and Lactobacillus casei. Certain strains of these species require vitamin B_6, but they can grow without this vitamin in the presence of *d*alanine (the dextro-rotatory, naturally occurring form of the amino acid). In this instance, however, the amino acid is not the precursor of the vitamin, but apparently the vitamin is necessary in order to produce the essential amino acid.

These cases, interesting in themselves, indicate also that the need of a particular organism for a given vitamin may to some extent depend on the diet. As a result of our knowledge concerning the chemical nature of the vitamins, many of them can be synthesized and are actually being manufactured on a large scale. We also have much knowledge concerning the specific vitamin requirements of many higher animals, of the isolated roots of green plants, of bacteria, fungi and protozoa. More and more information is becoming available concerning the vitamin needs of various insects. Vitamin studies are relatively simple for terrestrial forms-mammals, birds, insects. Aquatic forms are difficult material.

Perhaps if fish could be grown in sterilized water and kept sterile, correct data on their vitamin requirements could be obtained. With ordinary methods of attack, one can never be certain of the contribution that bacteria and other microorganisms are making to the vitamin content of the food taken in by the fish. Even in terrestrial animals, microorganisms may play a part in providing vitamins.

The bacteria in the intestinal tract can and do manufacture vitamins. In experiments with rats, it is sometimes necessary to prevent the animals from eating their feces, for such fecal material may be a source of vitamins. In many instances a mammal may not need a particular vitamin as food because it is manufactured by the bacteria of its intestine.

By giving the animal antibiotics such as sulfa drugs and thus eliminating or reducing its intestinal bacterial flora, it is possible to investigate the need for a vitamin such as folic acid. The nutritional requirements of various types of animals may depend largely on the parasites or symbionts they harbour.

Many species of insects, especially among the Hemiptera, have a constant association with symbiotic fungi. Even protozoa may contain symbionts or parasites within their single-celled bodies. Most of the work on vitamins has been stimulated by the practical needs of medicine and industry. The physician and the drug manufacturer are not ordinarily interested in the vitamin requirements of lower organisms, unless these organisms can conveniently be used in vitamin assay. Thus, as might be expected, there are vast gaps in our knowledge of the vitamin requirements of many large groups of animals. We know almost nothing of the vitamin needs of metazoan invertebrates other than insects.

And we know very little about the vitamins necessary for fish, amphibia and reptiles. A wide field is open for such study, and many facts of interest may well be discovered. Some vitamins are required by practically all living organisms from bacteria to man; others are needed by only some forms and not others. There has been much interest in the avitaminoses, that is to say the diseases caused by the lack of specific vitamins.

Our knowledge in this field is almost entirely confined to higher animals. Thus we know that when the vitamins D are omitted from the diet, a child will develop rickets and so will a pig or a rabbit, a dog or a chicken. Accordingly, the vitamins D have been called the antirickets vitamins. But animals without bones such as insects or protozoa may also require vitamin D. From a knowledge of the reasons why this is so, we may be able to obtain additional information as to the way that vitamins function.

Much as is known about the vitamins, their chemistry, their occurrence in various foods, the diseases that follow their absence from the diet, there is still almost complete ignorance as to the exact nature of their action. A beginning has been made in the recognition of the fact that some vitamins such as riboflavin, nicotinic acid, and pantothenic acid form essential parts of some enzyme systems.

We know also the end results of many types of vitamin deficiency. But to say that a vitamin is necessary for growth or that it prevents anemia tells us very little as to what it really does or why it is necessary. If we were to know how certain vitamins influence growth or prevent cell deterioration in the epithelial or nervous system, we would have basic information of great importance. Such information would be useful not only in the better understanding of the vitamins but also in the elucidation of various basic vital processes. The final interpretation of vitamin action must lie in an understanding of how the vitamins affect the cell protoplasm.

It is this aspect of the subject that will be stressed in the following discussion of individual vitamins. The general physiologist is not primarily interested in the relation of vitamins to human welfare, the foods a man should eat, or the benefits to be derived from adding this or that vitamin to the diet. Such information can be obtained in books on physiological chemistry, nutrition, or medicine. In discussing the individual vitamins, we shall not proceed in alphabetical order. This order has no other basis than the chronology of discovery. We shall consider first the fat-soluble and then the water-soluble vitamins.

In general, especially in the case of the water-soluble vitamins, true chemical names have tended to supplant letters and subscripts. But in some instances, several chemical substances can and normally do supply a given vitamin need. In these instances the alphabetical name is sometimes more convenient.

VITAMIN A

Vitamin A is a yellow viscous oil. Chemically, it is related to carotene, the common yellow pigment of plants. The generally accepted formula follows:

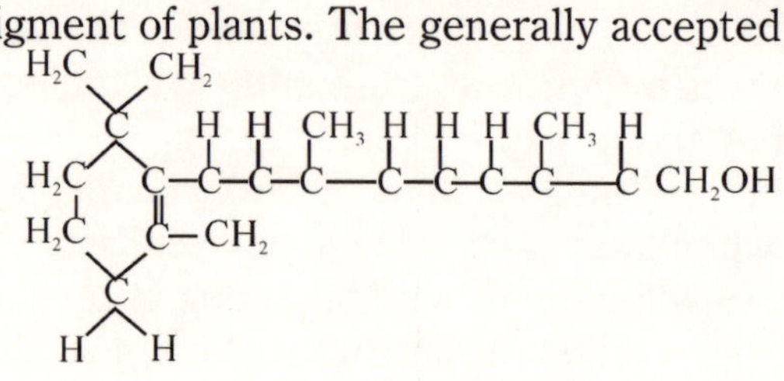

When carotene is eaten, it is converted to vitamin A. Because of this, it is sometimes called a provitamin. Another provitamin of vitamin A is kitol, a dihydric alcohol found in whale liver oil. The conversion of carotene to vitamin A occurs in the intestines of higher animals. The vitamin is then stored in the liver. The livers of polar bears and Arctic foxes are so rich in vitamin A that they are toxic. The toxicity is an example of what is known as hypervitaminosis.

Vitamin A has been synthesized in several different laboratories. There are really two forms of the vitamin. One form (A1) constitutes most of the vitamin content of the livers of salt water fishes. In the livers of fresh water fishes vitamin A_2 occurs. The 2 substances give slightly different colours with antimony trichloride and can be distinguished spectro-scopically.

In higher animals, vitamin A is important for growth. When rats are deprived of vitamin A, the cornea of the eye becomes horny. This condition is known as xerophthalmia. It can occur also in children. Not only the corneal epithelium, but other types of epithelia are also adversely affected in vitamin A deficiency; this may lead to lessened resistance to infection. Vitamin A has a relation to vision. In man, vitamin A deficiency causes night blindness, but the effect is not as readily produced as formerly supposed.

In so far as they have been investigated, invertebrates, with the apparent exception of the snail Helix, do not seem to require vitamin A. Aside from its relation to vision, little is known as to the reason why vitamin A is necessary. Its presence in cells can be recognized with a fluorescence microscope because of the fact that in ultraviolet light it gives a green fluorescence.

STEROLS-VITAMIN D

A number of sterols have vitamin D activity, that is to say they prevent the occurrence of rickets in higher animals. The chemistry of the sterols is very complicated, and books on physiological chemistry may be consulted for structural formulae, etc. When the common sterol of fungi, ergosterol, is irradiated with ultraviolet rays, another sterol, calciferol, is produced. This is commonly referred to as vitamin D_2. (There is no D_1, for the original use of the term was a misnomer). D_3 is 7-dehydrocholesterol.

Numerous other antirachitic (rickets-preventing) substances have also been described. Whether or not a given sterol or sterol derivative has antirachitic action can only be determined by actual test. Bacteria do not seem to require sterols.

Some species of protozoan flagellates do; also Entamoeba histolytica, the ameba that causes human dysentery. Insects, in so far as they have been studied, require sterols, but they seem to need cholesterol, a cholesterol derivative, or a plant sterol; calciferol cannot usually be substituted for these. In mammals and birds, vitamin D causes an increase in the deposition of calcium and phosphorus in the bones.

This is apparently due, in part at least, to an increased absorption through the intestine. Using radioactive isotopes, Greenberg found that vitamin D promotes the absorption of calcium and strontium from the digestive tract. Sterols may act in the same way in promoting the passage of calcium through the walls of the insect intestine. Moreover, they may be concerned with the entrance of calcium into cells. This aspect of the subject has never been investigated.

VITAMIN E

A third fat-soluble vitamin or group of vitamins is vitamin E, or perhaps one should say the vitamins E. There are very many substances with vitamin E activity, but the term vitamin E is restricted to the tocopherols. (The word tocopherol is derived from Greek words signifying child birth). There are 4 tocopherols now known: alpha, beta, gamma and delta tocopherol. The formula for alpha tocopherol is:

Alpha Tocopherol

Beta and gamma tocopherol have 1 less methyl group than alpha tocopherol and delta tocopherol has 2 less methyl groups. Vitamin E is necessary for the reproduction of the rat and other mammals and birds. If the male rat is fed a diet lacking in vitamin E, the germ cells in the testis degenerate. (The effect is not evident in the mouse, but is said to occur in the little fish commonly known as the guppy). In the absence of vitamin E, the female rat cannot give birth to its young; these degenerate and are resorbed. In some cases, muscular atrophy follows lack of vitamin E in the diet. In 1949, Harris wrote: "The mechanism whereby vitamin E functions in the body is still unknown."

A dual role was proposed by Hickman and Harris. In the first place it may act as part of an enzyme system, and to this action its effect in preventing degeneration of muscle may be due. Secondly, the tocopherols exert another, less specific function in that they tend to prevent excessive oxidation. Thus, for example, they tend to preserve vitamin A and they may also prevent the necessary unsaturated fats from being oxidized.

VITAMIN K

Fourth and last of the commonly recognized fat-soluble vitamins are the vitamins K, so named because they are important for blood coagulation (coagulation is spelled with a K in German and in the Scandinavian languages). The K vitamins are close to the E vitamins chemically. Tocopherols are very similar to the naphthoquinones chemically; the K vitamins are actually naphthoquinones. This is a 1,4 naphthoquinone.

The most active form of vitamin K, a form which can be prepared synthetically, is 2-methyl-1,4-naphthoquinone. This is sometimes called menadione. It does not occur naturally nearly as abundantly as vitamin K1, which resembles menadione except that instead of having a methyl group in the 3 position, it has a phytol group.

(Phytol is a 20 carbon mono-unsaturated primary alcohol which can be obtained from the saponification of chlorophyll). Vitamin K_2 also resembles menadione except that in the 3 position it has a difarnesyl group. (Farnesol is similar to phytol; it is a 15 carbon primary alcohol with 3 unsaturated carbon atoms). Several other naturally occurring vitamin K compounds have also been described. They are all either yellow oils or yellow crystalline solids, practically insoluble in water. Vitamin K is manufactured in one form or another by bacteria, which do not ordinarily require it.

The bacteria in the intestines of mammals produce enough so that normally the mammals have no need of it from outside sources. However, when absorption of fatty compounds is interfered with, as in intestinal disease or in liver disease involving interruption of bile flow, the vitamin becomes essential for man and mammals. Many birds require the vitamin, for their bacteria can not supply them with sufficient quantity. Vitamin K is necessary for the clotting of the blood of higher animals. The reason for this is not clearly understood.

To some extent opinions differ, but most authorities believe that this action of vitamin K is an indirect effect caused by the entrance into the blood of greater amounts of a substance prothrombin, the precursor of thrombin. Thrombin, which is regarded as an enzyme, is of primary importance in blood clotting.

Certainly the function of vitamin K can not be restricted to its effect on the clotting of blood. Obviously, the bacteria and the plants which produce it must have some need for it themselves. Experiments with isolated cells show that vitamin K is a very powerful protoplasmic clotting agent.

This is not surprising in view of the fact that the clotting of protoplasm is in so many respects similar to the clotting of blood. Although menadione is but sparingly soluble in water, solutions of it cause a coagulation of the protoplasm of marine eggs. Menadione also produces clotting in muscle protoplasm.

ASCORBIC ACID (VITAMIN C)

The fat-soluble vitamins are still generally referred to by their alphabetical names. On the other hand, the water-soluble vitamins are now for the most part usually given chemical names which have no relation to the alphabet. This became necessary when what was originally the B vitamin became a whole assortment of heterogeneous substances, distinguished by somewhat variable and uncertain subscripts of the letter B. At present the water-soluble vitamins include the clearly defined ascorbic acid (vitamin C) plus the vitamins of the B complex.

Because of the fact that vitamin C is one substance, clearly defined and understood, we shall consider it first rather than in its alphabetical order.

Ascorbic acid is now well understood chemically and it can indeed be synthesized. The L-form is much more potent than the D-form. Ascorbic acid is really an acid hexose sugar.

Its acidity is due to the dissociation of an enolic hydrogen rather than to opening of the lactone ring. Ascorbic acid is a strongly reducing substance, and because of this, it is easy to demonstrate its presence in cells and tissues. A standard method involves the reduction of silver nitrate. Because of its reducing power, there is doubtless a relation of ascorbic acid to the oxidative processes in cells, but there is still no clear understanding as to the exact function or functions of the vitamin in metabolism.

In animals suffering from scurvy there is lack of intercellular material; presumably the cells enclosing the blood vessels are not held together as firmly as they should be and this is doubtless the cause of hemorrhage. Mucopolysaccharides are known to be important constituents of the cementing substances which hold cells together. In view of the fact that mucopolysaccharides are polymers of acid sugars, it is possible that ascorbic acid is involved in the synthesis of such substances.

There is some indirect evidence for this. Normally in wound healing mucopolysaccharides are detectable in the healing tissues and may have considerable importance for the process. However, when the store of ascorbic acid is depleted, these substances do not appear.

THIAMINE (ANEURINE, VITAMIN B1)

The formula for thiamine is $C_{12}H_{18}N_4OS$. It is prepared synthetically as the chloride hydrochloride. This is a white crystalline compound readily soluble in water. Thiamine is a combination of a pyrimidine ring and a sulfur-containing thiazole ring.

Apparently all sorts of protoplasm need thiamine. Some types of cells can synthesize it, others can not. Higher organisms generally require the intact thiamine molecule; on the other hand some lower organisms are able to synthesize thiamine provided they can obtain both parts of the molecule, pyrimidine and thiazole. Other organisms may require either the pyrimidine fraction or the thiazole fraction, but not both. Many bacteria require neither and can synthesize whatever thiamine they need.

Lack of thiamine in the diet produces beriberi, a paralytic disease of men rather common in Japan and in various other portions of the world.

In birds, lack of thiamine produces polyneuritis, a disease similar to beriberi. In this disease the nerve fibres show degenerative changes. On a fox ranch in Minnesota owned by a Mr. Chastek, it was found that the inclusion of raw fish in the food given the foxes caused a paralysis now known as Chastek paralysis.

Experimentally, it has been shown that a similar paralysis could be produced in chicks, cats, pigeons and rats. The effect is due to the fact that some species of fish contain in their tissues an enzyme, thiaminase, which destroys thiamine. Thiaminase is found not only in fish but also in clams. In addition to its effect on the nervous system, lack of thiamine may also cause retardation of growth. Thiamine is important for the utilization of carbohydrate. Combined with 2 phosphate groups as diphosphothiamine, it forms part of an enzyme system that acts on pyruvic acid, a keto acid of great importance in intermediary metabolism. When animals are deprived of thiamine, their tissues are unable to oxidize pyruvic acid.

According to von Muralt, thiamine is also important in the transmission of the nerve impulse. When isolated nerves are stimulated, they are believed to release thiamine to the surrounding solution.

RIBOFLAVIN

Riboflavin is an orange crystalline substance, sparingly soluble in water, and markedly sensitive to light. The riboflavin molecule is complex. For an understanding of its structure and its synthesis as first accomplished by Karrer and his associates in 1935, the larger textbooks of organic chemistry may be consulted. The three rings represent an isoalloxazine constituent and this is combined with ribose, the pentose sugar also present in ribonucleic acids. Although riboflavin is apparently a constituent of all or almost all types of protoplasm, many lower organisms do not require it in their food.

Yeasts and molds have no need for it. Bacteria, in general, can get along without it. However, riboflavin serves as a growth factor for the tetanus bacillus, for lactic acid and propionic acid bacteria, and it is a requirement for Erysipelothrix rhusiopathiae and Listerella monocytogenes. The protozoan ciliates that have been investigated require riboflavin. So, too, do almost all the insects that have been studied. Riboflavin is essential for the growth and normal health of the rat, mouse, dog, pig, man, chicken and turkey. In riboflavin deficiency, various symptoms of ill health may appear.

These include redness and roughness of the cornea of the eye, soreness at the corners of the lips, redness and soreness of the tongue, etc. At one time, it was thought that "ariboflavinosis," the disease caused by riboflavin deficiency, was extremely common. But the symptoms mentioned above are not always due to lack of riboflavin. The presence of riboflavin in cells may be detected by fluorescence microscopy. Riboflavin is important for cell respiration. It and other alloxazine compounds combine with phosphoric acid and protein to form enzyme systems known as flavoproteins.

NICOTINIC ACID (NIACIN)

Nicotinic acid and pyridoxine (to be considered next) are closely related substances. Both are derivatives of pyridine. This substance is a weak base, the molecule of which contains five CH-groups and a nitrogen atom arranged in a six-membered ring. As is customary with ring compounds, the various positions on the ring have numbers. In pyridine, nitrogen occupies the number 1 position. Nicotinic acid is 3-pyridine carboxylic acid. The structural formulas of the acid and its amide are given below:

The amide is more important than the acid itself, for some organisms require it and cannot manufacture it from the acid. Both acid and amide are white crystalline powders, without odour. According to Stephenson, "Nicotinic acid is one of the commonest growth factors needed by bacteria; in other words the ability to synthesize it is one most frequently lost. This may be due to its very wide distribution in soil and natural media." It is an essential requirement for those ciliate protozoa that have been studied (Colpoda and Tetrahymena) and also for most of the insects that have been investigated. Nicotinic acid is essential for the dog, the pig, and for man. However, if enough tryptophan is added to the diet, neither dog nor pig requites it.

As noted previously, tryptophan is a precursor of nicotinic acid, and the tissues can synthesize the vitamin from tryptophan. From a lack of nicotinic acid, dogs become afflicted with what is known as blacktongue disease. In man, nicotinic acid deficiency is the primary cause of pellagra.

This serious disease, common enough in the United States, involves the lining of the digestive tract, the skin, and the nervous system. Nicotinic acid amide enters into the composition of two substances of great importance for the oxidative metabolism of the cell. These are coenzyme I and coenzyme II. Presumably nicotinic acid deficiency is associated with a lack in the tissues of coenzymes I and II. But what connection there is between such a lack and the manifold disturbances present in a disease like pellagra is not at all clear.

PYRIDOXINE (VITAMIN B6)

The term vitamin B6 is now properly used, according to a decision of the American Society of Biological Chemists, to designate three naturally occurring substances: pyridoxine, pyridoxal and pyridoxamine. Pyridoxal is the aldehyde derivative of pyridoxine and has a formyl group instead of a hydroxymethyl group at the 4 position. Pyridoxamine is an amino derivative and has an amino methyl instead of a hydroxymethyl group at the 4 position. The three forms of the vitamin are almost equally effective for higher animals, but some bacteria have specific requirements for one form or the other. Vitamin B_6 is needed by many bacteria, by some protozoa, and by most of the insects whose vitamin requirements have been investigated.

When rats are deprived of vitamin B_6, they develop a painful inflammation of the skin. This is especially evident in the skin of the extremities—the paws, snout, and ears. The ailment is sometimes called rat acrodynia because of a supposed similarity to a human ailment in which the nerves and skin of the hands and feet are affected. The disease is not wholly specific for vitamin B_6 deficiency; it can be produced also in other ways, for example by a lack of unsaturated fatty acids. In various higher animals, convulsions may follow extreme vitamin B_6 deficiency. Sometimes lack of the vitamin causes muscular weakness; it can also cause anemia.

Vitamin B_6 deficiency can be induced by the addition of an antagonistic analogue, desoxypyridoxine. Thus if this compound is injected into hens' eggs, the embryos do not proceed very far in their development. Vitamin B_6 is important for the proper metabolism of the amino acids. It forms part of an enzyme system which promotes the synthesis of amino acids by the addition of carboxyl groups. Acting as part of what is presumably the same enzyme system, the vitamin also favours the removal of carboxyl groups from amino acids.

Another type of enzyme for which vitamin B_6 is an important constituent is transaminase. This enzyme catalyzes the transfer of an amino group from one amino acid to another; an important type of reaction in protein metabolism. Thus there is some important information as to how the vitamin functions with respect to the chemical reactions occurring during the metabolism of proteins and amino acids, but how this is related to the disturbances which occur in skin, nerves, muscle and blood when the vitamin is lacking has yet to be explained. It has been claimed, on the basis of isolation studies of mitochondria, that these granules contain especially large amounts of vitamin B_6. However, it is the nucleoprotein-containing granules of the cytoplasm that are believed to be especially important for protein metabolism.

PANTOTHENIC ACID

Pantothenic acid is found in many diverse types of living materials. Chemically, it consists of β-alanine in a peptid linkage with a dihydroxy acid. (β-alanine differs from ordinary alanine in being β-aminopropionic acid instead of α-aminopropionic acid). The acid is unstable and for this reason is sold as a calcium salt. This salt is a white powder, readily soluble in water but not in alcohol. Many organisms from bacteria and protozoa to higher mammals require pantothenic acid. In its absence these organisms do not grow properly and they may show various ill effects.

Symptoms of pantothenic acid deficiency include dermatitis (inflammation of the skin) in rats, deficient feathering in chicks, graying of hair in rats, degeneration of the nervous system in pigs, hemorrhages in the adrenal gland and general disturbance of function of this gland, etc. The vitamin constitutes part of an enzyme system. Combined with various substances including phosphate, adenylic acid, and a sulfhydryl group, it is coenzyme A, and this in combination with a protein is an enzyme capable of acetylating various organic substances such as choline and sulfanilamide.

The acetylation of choline may be very important physiologically. How the acetylation, or rather the lack of it, can produce all the pathological conditions noted above is not clear, although in the case of the nervous system, any alteration of acetylcholine metabolism would certainly be harmful.

BIOTIN

As long ago as 1898 it was shown that if dogs were fed raw egg white, diarrhea resulted. Following this it was shown that other mammals also suffered from egg white feeding, and that in addition to the diarrhea, skin ailments and nervous symptoms appeared. The phenomenon became known as egg white injury. Cooking the egg white prevented the injury. György proposed the term vitamin H for a factor which prevented egg white injury. Later, it was shown that this factor is identical with biotin, a vitamin known to be important for yeasts and microorganisms.

The raw egg white unites with the biotin and prevents its utilization by mammals or birds. Chemically, biotin is rather complicated. It has in its molecule a urea ring containing one sulfur atom; the ring is combined with a valeric acid group. The presence of sulfur in the molecule is not important for the vitamin activity, for if the sulfur is replaced by oxygen, the vitamin activity is retained. Biotin acts in extremely low concentration, and it is thus one of the most potent of the vitamins. Many organisms do not grow unless biotin is available for them. Usually in higher organisms it is synthesized in sufficient quantity, but biotin deficiency can occur naturally in birds and in insects.

Lack of sufficient biotin has the same effect as egg white injury—diarrhea, disturbances in the skin and nervous system, in rats loss of hair about the eyes. It can also prevent the birth of young in rats.

The protein in egg white which binds biotin is called avidin; it is a basic, carbohydrate-containing protein and it occurs not only in the eggs of birds but also in frog eggs. There is now strong evidence that biotin is important in a process which is known as carbon dioxide fixation. Many bacteria, and higher organisms also, are able to utilize carbon dioxide and form organic compounds from it. This process is interfered with if biotin is lacking.

FOLIC ACID

This vitamin occurs in green leaves; hence the name, which is taken from the Latin word for leaf. Folic acid also occurs in mushrooms and yeast; and in animal tissues, such as liver and kidney. The vitamin has been called by many different names, and it exists in various forms. These have different potencies for different types of organisms.

Chemically, folic acid consists of pteroic acid combined with the common amino acid, glutamic acid. Pteroic acid itself is a combination of para-aminobenzoic acid and a pteridine base. Pteridine bases are pyrimidine derivatives related to the pyrimidine bases thymine and uracil found in the nucleic acids.

The compound shown above is pteroylglutamic acid, sometimes called PGA. Pteroic acid may also combine with 3 glutamic acid groups to form pteroyltriglutarnic acid (P3GA) or with 7 molecules of glutamic acid to form pteroylh-eptaglutamic acid (P7GA). All of these compounds are found naturally and all have vitamin activity for certain types of living materials. Folic acid is required by some bacteria, by certain ciliate protozoal, by some insects and by various birds and mammals. The bacteria in the rat intestine normally manufacture a sufficient quantity of the vitamin, and treatment with antibiotics such as sulfa drugs is necessary to cause folic acid deficiency.

Folic acid is commonly an essential growth factor, not only for bacteria but for higher organisms as well. Lack of the vitamin in mammals results in various ailments, but especially in a type of anemia characterized by a small number of red blood cells of large size. This type of anemia, called macrocytic anemia, occurs in the tropical disease sprue, and sprue can be cured either by liver extract (which contains folic acid) or by folic acid itself. Folic acid is apparently important both for growth and cell division. When chick embryos are in contact with antagonistic analogues such as 4amino folic acid, their growth is greatly retarded.

Similar results have been obtained in numerous other studies with folic acid antagonists. Thus 4-amino folic acid has a retarding effect on tumor growth. A substance related to folic acid is the so-called "citrovorum factor," which is apparently identical with "folinic acid." This factor favours growth of the bacterium, Leuconostoc citrovorum. The citrovorum factor may be obtained from liver extracts; also it is excreted into the urine of rats or men fed folic acid.

VITAMIN B12

Year by year, the number of B vitamins increases. B_{10} and B_{11} are required by chicks, the former for proper feathering, the latter for growth. From liver it has been possible to isolate a pink crystalline substance which can be used to cure pernicious anemia.

Folic acid is useful in this disease, for it has a hematopoietic effect (that is to say, it favours increased production of blood cells), but it does not correct the nervous symptoms of the disease. These depend on degenerative changes in the nervous system. Vitamin B_{12} in tiny amounts has been thought to correct diseases of both blood and nervous systems, but apparently if the nervous system is badly affected, it can not be cured.

The vitamin has been crystallized. It contains 4-4.5 per cent cobalt and also phosphorus; the molecular weight is approximately 1500. (Estimates vary from 1300 to 1600). In liver extracts there are really two active substances with slightly different absorption spectra. One of these is the definitive B_{12}, the other is B_{12b}. (B_{12a} is another similar substance derived from B_{12} by hydrogenation). Vitamin B_{12} is apparently identical with what has been called "animal protein factor." This was originally obtained from such diverse sources as milk, muscle, liver and cow manure. Although called animal protein factor, it can likewise be obtained from plant sources.

Both animal protein factor and vitamin B_{12} promote growth in some bacteria, in chickens, rats and mice. The effect on growth may be due to the fact that the vitamin seems to increase the utilization of amino acids. At any rate, this has been reported for chicks.

OTHER POSSIBLE VITAMINS OF THE B COMPLEX

B_{13} has been described as a growth factor for rats and for pigs. There is also a description of a vitamin B_{14}, extracted from urine. Over the years, there has been a large and controversial literature about what has been called vitamin P. Substances with vitamin P activity are supposed to maintain proper permeability of the blood capillaries of higher animals and to prevent excessive loss of fluid from them. (The letter P is due to this permeability action). There is now general agreement that substances with P activity are not to be considered as vitamins.

Para-aminobenzoic acid (often called paba). This relatively simple compound is an important growth factor for many bacteria. Thus sulfanilamide is an antagonistic analogue of para-aminobenzoic acid, and its antibacterial action is due to this similarity in structure. The discovery of this relationship in 1940 was indeed a milestone in our understanding of the mechanism of drug action and has served as a model and guide for many subsequent researches.

Choline

This substance, the basic constituent of lecithin, is believed by many to be a vitamin. Rats deprived of choline develop fatty livers and other types of hepatic injury. They may also show degenerative changes in the kidney. Choline deficiency is also a factor in the development of a leg deficiency (perosis) in chicks and turkeys.

It should be remembered that choline is a constituent not only of lecithin, but also of the important humoural substance, acetylcholine. For a discussion of the significance of acetylcholine, *Inositol*. Inositol is a derivative of benzene.

It can be prepared from hexahydroxy-benzene by reduction, in the course of which 6 hydrogen atoms are added. Thus inositol is hexahydroxy-hexahydro-benzene. It is an optically active substance found in yeast, muscle, and in various other types of animal and plant tissue. Inositol is important for the growth of yeasts, and is indispensable for some types of yeast.

It also promotes the growth of some fungi. In the presence of inositol, chicks grow more rapidly than when it is absent. However, rats may be bred for three generations without inositol, and György was led to conclude that "the facts presented do not, at the present time, warrant the identification of inositol as a primary and really essential vitamin; they rather favour the assumption that it is a supporting and at least not always specific vitamin." *α-Lipoic Acid*. This substance, also called protogen, is probably to be included among the vitamins of the B group, for it is water soluble.

It is important for the growth of some bacteria and for the protozoan Tetrahymena, and it is involved in the oxidation of the important keto acid, pyruvic acid (for this reason it was formerly called pyruvate oxidase factor). Many types of living material contain lipoic acid. It is now known to be a derivative of octanoic (caprylic) acid and it contains sulfur in the form of an S-S group. There are doubtless other vitamins of the B group still to be discovered. Fraenkel found that the larva of the beetle Tenebrio molitor (the so-called mealworm) requires some substance present in yeast or liver but not one of the known B vitamins. This he called B_T. More recently he has identified it with carnitine, a dipeptid known originally in muscle.

Also the flour beetle requires at least two other unidentified members of the B complex and neither of these is B_T. Goetsch claims that there is a vitamin present in certain insects, especially termites, a vitamin which has pronounced effects on growth and differentiation. This he calls the vitamin T complex; it is supposed to affect not only insects but various other types of organisms. Throughout the course of vitamin study, there have indeed been many extravagant claims and many false leads. The vast literature is often contradictory and confused. But in spite of this, in the course of time the truth has gradually emerged and the main outlines of the picture are now clear.

All sorts of organisms, from bacteria to man, require certain specific types of organic molecules. These can be synthesized by green plants and by many lower organisms. Higher forms have frequently lost the power for such synthesis, and unless, like the ruminants, they provide themselves with huge storehouses for bacteria, they must depend for their existence on obtaining the vitamins with their food. From the knowledge of the nature of the vitamins, we can begin to develop a chemical anatomy of living systems. The vitamins represent indispensable organic molecules and apparently the same molecules are almost universally needed. Moreover, we are beginning to discover why the vitamins are needed and what work they do in the living cell. Much of our information is still fragmentary.

The vitamin requirements for the lower vertebrates and for the hosts of invertebrate animals have scarcely been investigated at all. Moreover, in some instances we are completely at a loss to account for the fundamental necessity for a given vitamin.

Such superficial disturbances as a reddening of the eye or a scaliness of the skin give no direct indication of the part a vitamin may play in protoplasmic activity.

3

Nutrient Requirements: Cow and Rumen

REQUIRED NUTRIENTS

All living organisms require essential nutrients to support metabolic processes to keep them alive. General classification of required nutrients include: water, the most essential, energy, protein, minerals, and vitamins. Minerals can be further subdivided into macrominerals, microminerals based on the daily amounts (gm or mg) required. Vitamins are separated into fat or water soluble sources.

Daily requirements for these essential nutrients are a function of the cow's body weight and physiologic state (*e.g.*, maintenance, growth, lactation, pregnancy) as modified by environmental conditions. Bacteria have similar requirements for maintenance and growth (*i.e.*, reproduction). Differences between the cow and microbes are seen in where they derive their nutrients. As a consequence of a pregastric fermentation system, consumed feeds are exposed to microbial fermentation before being available for digestion and absorption by the host.

In many instances, this process enhances nutrient availability to the host animal. However, an opposite effect can also occur. Exposure to microbial fermentation can result in nutrient alteration to a form that is either no longer absorbed or has lost its biologic activity.

As an anaerobic fermentation system, the rumen is a highly reduced environment with an abundance of reducing equivalents looking for a place to go. Highly oxidized compounds are prime targets. Absorption of selenium (Se) occurs only in its highly oxidized forms as selenite (+4) or selenate (+6) via an intestinal sulfate transport system. Selenium absorption from selenite, is more efficient in non-ruminants than ruminants with a total retention of 77 per cent and 29 per cent, respectively. Microbial reduction of Se to its elemental (+0) or selenide (-2) forms renders it unavailable thus reducing Se absorption in ruminants. High concentrate diets are associated with increases in insoluble Se complexes, possibly due to changes in rumen pH, redox potential, microbial populations or some combination.

Table. Substances Which Supply Essential Nutrient Needs for the Cow and Rumen Microbial Population.

Nutrient	Cow	Bacteria
Energy	VFA's Glucose	Complex Carbohydrates Sugars, Starches, Amino Acids
Protein	Amino Acids Microbial Protein	Ammonia, Amino Acids, Peptides
Minerals	Dietary	Dietary
Vitamins	Dietary Bacterial	Dietary Synthesized

Rumen interactions can also result in a reduced availability of other minerals, namely copper (Cu) and magnesium (Mg). Best studied relative to copper availability are the interaction of molybdenum (Mo) and sulfate (SO_4) in the rumen. High Mo and SO_4 concentrations in the rumen allow microbial synthesis of thiomolybdates, which can chelate copper making it unavailable for absorption or utilization.

Dietary Mg absorption efficiency is low and occurs in the rumen by a sodium (Na)-potassium (K) ATPase system. Many dietary factors, including excess K, nitrogen, Ca, sulfur and organic acids as well as deficient Mg and phosphorus levels, can predispose an animal to a form of hypomagnesemic tetany. Another factor, a direct consequence of rumen microbial activity, also plays an important role in the pathogenesis of this syndrome. Grasses typically contain significant amounts (>1 per cent of plant dry matter) of sucrose and trans-aconitate, an organic acid, during periods of rapid growth. In the rumen trans-aconitate is metabolized to acetate or reduced to tricarballyic acid. One of the most abundant rumen bacteria, *Selenomonas ruminantium*, is primarily responsible for the reduction of trans-aconitate. This organism also prefers to grow when sucrose is readily available as substrate. Tricarballylic acid is a very potent chelator of calcium (Ca), Mg and Zn inducing increased urinary excretion and reduced tissue status of these minerals.

Making more bugs - Microbial Protein

The cow derives a majority of her energy and protein from microbial end products. In other words, the more we make the bugs grow (reproduce), the less additional, more expensive feedstuffs we need to provide in the cow's diet. Why is production of microbes so important to a ruminant feeding programme? Microbes contain approximately 62 per cent crude protein, which is 80 per cent true protein and 80 per cent digestible. This is considered high quality protein. Microbial protein production alone can support up to 25 kg of milk production.

The first goal of a ruminant feeding programme should be to maximize microbial protein production and then secondly, meet additional cow's nutrient requirements over-and-above those not met by microbial fermentation end products. This type of feeding approach would theoretically be the most economical and efficient. So how do we get the rumen microbes to abundantly grow without disturbing the rumen ecosystem?

Table. Microbial Protein Synthesis Relative to Daily Protein Needs of the Cow.

Efficiency of Microbial Protein Synthesis	**Daily Milk Yield**		
	25 kg	**35 kg**	**45 kg**
gm N/kg OM digested	% of protein from microbes		
20	49	42	39
30	73	64	59
40	98	85	79

Bacteria require a number of essential nutrients for the synthesis of protein, similar to that of the cow. However unlike the cow, bacteria can use a greater variety of potential nitrogen sources to synthesize amino acids, the building blocks of proteins. In addition, bacteria can synthesize both essential and non-essential amino acids unlike the cow which needs to be supplied with preformed essential amino acids. Figure presents an overview of the processes required to synthesize microbial protein.

The microbial protein production is a function of rumen available substrates, primarily carbohydrates and nitrogen. If any of the required building blocks are in limited supply, microbial protein production will be determined by the availability of the most limiting substrate. Usually this is energy from carbohydrate fermentation.

Energy production (generation of ATP) will be dependent upon the available carbohydrate source and its rate of degradation. Ammonia (NH_3) may be provided from non-protein nitrogen sources, amino acids, peptides, or proteins where utilization of a nitrogen source is dependent upon the specific population of bacteria. For example, cellulolytic bacteria can only use NH_3 as their nitrogen source.

Microbial protein production is more complex than just providing the necessary amounts of substrate in the diet. The rumen is a dynamic system that constantly has fermentation end products, liquid, bacteria, and particles being removed via digestion and passage through the rumen as well as new substrate added. So not only do we need to address concepts of total substrate requirements, availability of substrate relative to other substrates needs to be addressed. We must be able to,

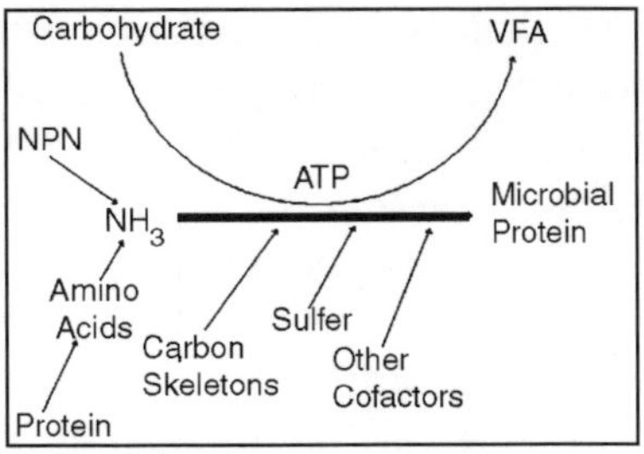

Metabolic processes involved in microbial protein synthesis.

NPN = non-protein nitrogen; NH_3 = ammonia.

Predict rate and extent of carbohydrate and protein degradation takes place in the rumen. This is the critical component of a dynamic modeling system for the rumen and requires more comprehensive and complex feed analysis procedures.

ANIMAL NUTRITION

Nutritional science investigates the metabolic and physiological responses of the body to diet. With advances in the fields of molecular biology, biochemistry, and genetics, the study of nutrition is increasingly concerned with metabolism and metabolic pathways: the sequences of biochemical steps through which substances in living things change from one form to another. Carnivore and herbivore diets are contrasting, with basic nitrogen and carbon proportions being at varying levels in particular foods. Carnivores consume more nitrogen than carbon while herbivores consume less nitrogen than carbon, when an equal quantity is measured. Nitrogen is the predominant source of energy for organisms, through the nitrogen cycle, with various minerals and vitamins being equally vital. Nitrogen is also the primary energy for regular metabolic rates, this necessitates greater sums of plants for herbivore digestion, to absorb significant energy for sustenance and stamina.

The human body contains chemical compounds, such as water, carbohydrates, amino acids, fatty acids, and nucleic acids. These compounds in turn consist of elements such as carbon, hydrogen, oxygen, nitrogen, phosphorus, calcium, iron, zinc, magnesium, manganese, and so on. All of these chemical compounds and elements occur in various forms and combinations, both in the human body and in the plant and animal organisms that humans eat.

The human body consists of elements and compounds ingested, digested, absorbed, and circulated through the bloodstream to feed the cells of the body. Except in the unborn fetus, the digestive system is the first system involved. In a typical adult, about seven litres of digestive juices enter the lumen of the digestive tract.

These digestive juices break chemical bonds in ingested molecules, and modulate their conformations and energy states. Though some molecules are absorbed into the bloodstream unchanged, digestive processes release them from the matrix of foods. Unabsorbed matter, along with some waste products of metabolism, is eliminated from the body in the feces.

Studies of nutritional status must take into account the state of the body before and after experiments, as well as the chemical composition of the whole diet and of all material excreted and eliminated from the body. Comparing the food to the waste can help determine the specific compounds and elements absorbed and metabolized in the body.

The effects of nutrients may only be discernible over an extended period, during which all food and waste must be analysed. The number of variables involved in such experiments is high, making nutritional studies time-consuming and expensive, which explains why the science of human nutrition is still slowly evolving.

In general, eating a wide variety of fresh, whole, foods has proven favourable for one's health compared to monotonous diets based on processed foods. In particular, the consumption of whole-plant foods slows digestion and allows better absorption, and a more favourable balance of essential nutrients per Calorie, resulting in better management of cell growth, maintenance, and mitosis, as well as better regulation of appetite and blood sugar. Regularly scheduled meals have also proven more wholesome than infrequent or haphazard ones, although a recent study has also linked more frequent meals with a higher risk of colon cancer in men.

CARBOHYDRATES

Carbohydrates include sugars, starches and fibre. They constitute a large part of foods such as rice, noodles, bread, and other grain-based products. Carbohydrates may be classified chemically as monosaccharides, disaccharides, or polysaccharides depending on the number of monomer units they contain. Monosaccharides, disaccharides, and polysaccharides contain one, two, and three or more sugar units, respectively.

Polysaccharides are often referred to as complex carbohydrates because they consist of long, sometimes branched chains of single sugar units. Mono- and disaccharides are called simple carbohydrates. Dietary advice frequently but erroneously suggests that complex carbohydrates are superior to simple because they take longer to digest and absorb. Simple carbohydrates, on the other hand, are said to cause a spike in blood glucose levels rapidly after ingestion. These traditional claims are false In fact, many digestible polysaccharides are processed as rapidly and simple sugars in the human body. On the other hand some simple carbohydrates are processed in a different way and do not spike blood sugar. Thus the distinction between "complex" and "simple" does not predict the nutritional value or impact of carbohydrates. A better way of determining what effect particular foods may have on blood sugar and ultimately on health in general is the glycemic index.

Carbohydrates are not essential nutrients, but are typically an important part of the human diet. While it would not be accurate to categorize all carbohydrates as "bad" nutritionally, some carbohydrate sources may well have deleterious effects on health, especially when consumed in large quantities. Highly processed carbohydrates as well as fructose consumed in large quantities have been implicated in negative

Fibre

Dietary fibre is a carbohydrate that is incompletely absorbed in humans and in some animals. Like all carbohydrates, when it is metabolized it can produce four Calories of energy per gram. However, in most circumstances it accounts for less than that because of its limited absorption and digestibility. Dietary fibre consists mainly of cellulose, a large carbohydrate polymer that is indigestible because humans do not have the required enzymes to disassemble it. There are two subcategories: soluble and insoluble fibre.

Whole grains, fruits, and vegetables are good sources of dietary fibre. There are many health benefits of a high-fibre diet. Dietary fibre helps reduce the chance of gastrointestinal problems such as constipation and diarrhea by increasing the weight and size of stool and softening it. Insoluble fibre, found in whole-wheat flour, nuts and vegetables, especially stimulates peristalsis — the rhythmic muscular contractions of the intestines which move digesta along the digestive tract. Soluble fibre, found in oats, peas, beans, and many fruits, dissolves in water in the intestinal tract to produce a gel which slows the movement of food through the intestines. This may help lower blood glucose levels because it can slow the absorption of sugar. Additionally, fibre, perhaps especially that from whole grains, is thought to possibly help lessen insulin spikes, and therefore reduce the risk of type 2 diabetes. The link between increased fibre consumption and a decreased risk of colourectal cancer is still uncertain.

Fat

A molecule of dietary fat typically consists of several fatty acids, bonded to a glycerol. They are typically found as triglycerides. Fats may be classified as saturated or unsaturated depending on the detailed structure of the fatty acids involved. Saturated fats have all of the carbon atoms in their fatty acid chains bonded to hydrogen atoms, whereas unsaturated fats have some of these carbon atoms double-bonded, so their molecules have relatively fewer hydrogen atoms than a saturated fatty acid of the same length.

Unsaturated fats may be further classified as monouns-aturated or polyunsaturated. Furthermore, depending on the location of the double-bond in the fatty acid chain, unsaturated fatty acids are classified as omega-3 or omega-6 fatty acids. Trans fats are a type of unsaturated fat with trans-isomer bonds;

these are rare in nature and in foods from natural sources; they are typically created in an industrial process called hydrogenation. There are nine kilocalories in each gram of fat.

Saturated fats have been a staple in many world cultures for millennia. Unsaturated fats are considered healthier, while trans fats are to be avoided. Saturated and some trans fats are typically solid at room temperature, while unsaturated fats are typically liquids. Trans fats are very rare in nature, and have been shown to be highly detrimental to human health, but have properties useful in the food processing industry, such as rancidity resistance.

Essential Fatty Acids

Most fatty acids are non-essential, meaning the body can produce them as needed, generally from other fatty acids and always by expending energy to do so. However, in humans, at least two fatty acids are essential and must be included in the diet. An appropriate balance of essential fatty acids—omega-3 and omega-6 fatty acids—seems also important for health, although definitive experimental demonstration has been elusive.

Both of these "omega" long-chain polyunsaturated fatty acids are substrates for a class of eicosanoids known as prostaglandins, which have roles throughout the human body. They are hormones, in some respects. The omega-3 eicosapentaenoic acid, which can be made in the human body from the omega-3 essential fatty acid alpha-linolenic acid (LNA), or taken in through marine food sources, serves as a building block for series 3 prostaglandins.

The omega-6 dihomo-gamma-linolenic acid serves as a building block for series 1 prostaglandins, whereas arachidonic acid serves as a building block for series 2 prostaglandins. Both DGLA and AA can be made from the omega-6 linoleic acid in the human body, or can be taken in directly through food. An appropriately balanced intake of omega-3 and omega-6 partly determines the relative production of different prostaglandins, which is one reason why a balance between omega-3 and omega-6 is believed important for cardiovascular health. In industrialized societies, people typically consume large amounts of processed vegetable oils, which have reduced amounts of the essential fatty acids along with too much of omega-6 fatty acids relative to omega-3 fatty acids.

The conversion rate of omega-6 DGLA to AA largely determines the production of the prostaglandins PGE1 and PGE2. Omega-3 EPA prevents AA from being released from membranes, thereby skewing prostaglandin balance away from pro-inflammatory PGE2 towards anti-inflammatory PGE1. Moreover, the conversion of DGLA to AA is controlled by the enzyme delta-5-desaturase, which in turn is controlled by hormones such as insulin and glucagon. The amount and type of carbohydrates consumed, along with some types of amino acid, can influence processes involving insulin, glucagon, and other hormones;

therefore the ratio of omega-3 versus omega-6 has wide effects on general health, and specific effects on immune function and inflammation, and mitosis.

Iodolipids

Iodolipids are iodinated lipids which have important actions in biochemistry and physiology of all living cells. Iodine can add to double bonds of docosahexaenoic acid and arachidonic acid, making them less reactive to free oxygen radicals.. Multiple iodolipid classes comprising iodolactones and iodoaldehydes have been identified that have structural and metabolic functions in plant, animal and human cells.

Arachidonic acid and iodolactones specifically inhibit signal transduction pathways induced by local growth factors such as epidermal growth factor and basic fibroblast growth factor Delta-iodolactones, at physiological concentrations, have antiproliferative effects and could be intermediates in iodine-induced autoregulation of cell proliferation, especially of the thyroid gland and mammary gland.

Protein

Proteins are the basis of many animal body structures. They also form the enzymes that control chemical reactions throughout the body. Each molecule is composed of amino acids, which are characterized by inclusion of nitrogen and sometimes sulphur. The body requires amino acids to produce new proteins and to replace damaged proteins. As there is no protein or amino acid storage provision, amino acids must be present in the diet. Excess amino acids are discarded, typically in the urine. For all animals, some amino acids are essential and some are non-essential. About twenty amino acids are found in the human body, and about ten of these are essential and, therefore, must be included in the diet. A diet that contains adequate amounts of amino acids is particularly important in some situations: during early development and maturation, pregnancy, lactation, or injury. A complete protein source contains all the essential amino acids; an incomplete protein source lacks one or more of the essential amino acids. It is possible to combine two incomplete protein sources to make a complete protein source, and characteristic combinations are the basis of distinct cultural cooking traditions. Sources of dietary protein include meats, tofu and other soy-products, eggs, legumes, and dairy products such as milk and cheese. Excess amino acids from protein can be converted into glucose and used for fuel through a process called gluconeogenesis. The amino acids remaining after such conversion are discarded.

Minerals

Dietary minerals are the chemical elements required by living organisms, other than the four elements carbon, hydrogen, nitrogen, and oxygen that are present in nearly all organic molecules. The term "mineral" is archaic, since the intent is to describe simply the less common elements in the diet. Some are heavier than the four just including several metals, which often occur as ions in the body.

Some dietitians recommend that these be supplied from foods in which they occur naturally, or at least as complex compounds, or sometimes even from natural inorganic sources. Some minerals are absorbed much more readily in the ionic forms found in such sources. On the other hand, minerals are often artificially added to the diet as supplements; the most famous is likely iodine in iodized salt which prevents goiter.

MacromLnerals

Many elements are essential in relative quantity; they are usually called "bulk minerals". Some are structural, but many play a role as electrolytes.

Elements with recommended dietary allowance greater than 200 mg/day are, in alphabetical order:

- Calcium, a common electrolyte, but also needed structurally
- Chlorine as chloride ions; very common electrolyte
- Magnesium, required for processing ATP and related reactions
- Phosphorus, required component of bones; essential for energy processing
- Potassium, a very common electrolyte
- Sodium, a very common electrolyte; not generally found in dietary supplements, despite being needed in large quantities, because the ion is very common in food: typically as sodium chloride, or common salt. Excessive sodium consumption can deplete calcium and magnesium, leading to high blood pressure and osteoporosis.
- Sulphur, for three essential amino acids and therefore many proteins. Sulphur is not consumed alone, but in the form of sulphur-containing amino acids

Trace Minerals

Many elements are required in trace amounts, usually because they play a catalytic role in enzymes.

Some trace mineral elements are, in alphabetical order:

- Cobalt required for biosynthesis of vitamin B_{12} family of coenzymes. Animals cannot biosynthesize B_{12}, and must obtain this cobalt-containing vitamin in the diet
- Copper required component of many redox enzymes, including cytochrome c oxidase
- Chromium required for sugar metabolism
- Iodine required not only for the biosynthesis of thyroxine, but

probably, for other important organs as breast, stomach, salivary glands, thymus etc.; for this reason iodine is needed in larger quantities than others in this list, and sometimes classified with the macrominerals

- Iron required for many enzymes, and for hemoglobin and some other proteins
- Manganese
- Molybdenum required for xanthine oxidase and related oxidases
- Nickel present in urease
- Selenium required for peroxidase
- Vanadium
- Zinc required for several enzymes such as carboxypeptidase, liver alcohol dehydrogenase, and carbonic anhydrase

Vitamins

Some vitamins are recognized as essential nutrients, necessary in the diet for good health. Certain vitamin-like compounds that are recommended in the diet, such as carnitine, are thought useful for survival and health, but these are not "essential" dietary nutrients because the human body has some capacity to produce them from other compounds.

Moreover, thousands of different phytochemicals have recently been discovered in food, which may have desirable properties including antioxidant activity; however, experimental demonstration has been suggestive but inconclusive. Other essential nutrients that are not classified as vitamins include essential amino acids, choline, essential fatty acids.

Vitamin deficiencies may result in disease conditions, including goitre, scurvy, osteoporosis, impaired immune system, disorders of cell metabolism, certain forms of cancer, symptoms of premature aging, and poor psychological health, among many others. Excess levels of some vitamins are also dangerous to health, and for at least one vitamin, B6, toxicity begins at levels not far the required amount. Deficient or excess levels of minerals can also have serious health consequences.

Water

It is not fully clear how much water intake is needed by healthy people, although some assert that 6–8 glasses of water daily is the minimum to maintain proper hydration. The notion that a person should consume eight glasses of water per day cannot be traced to a credible scientific source. The effect of, greater or lesser, water intake on weight loss and on constipation is also still unclear. The original water intake recommendation in 1945 by the Food and Nutrition Board of the National Research Council read: "An ordinary standard for diverse persons is 1 millilitre for each calorie of food.

Most of this quantity is contained in prepared foods." The latest dietary reference intake report by the United States National Research Council recommended, generally,: 2.7 litres of water total for women and 3.7 litres for men. Specifically, pregnant and breastfeeding women need additional fluids to stay hydrated. The Institute of Medicine—who recommend that, on average, women consume 2.2 litres and men 3.0 litres—this is recommended to be 2.4 litres for pregnant women and 3 litres for breastfeeding women because an especially large amount of fluid is lost during nursing.

For those who have healthy kidneys, it is somewhat difficult to drink too much water, but it is dangerous to drink too little. People can drink far more water than necessary while exercising, however, putting them at risk of water intoxication, which can be fatal. In particular, large amounts of de-ionized water are dangerous. Normally, about 20 per cent of water intake comes in food, while the rest comes from drinking water and assorted beverages. Water is excreted from the body in multiple forms; including urine and feces, sweating, and by water vapour in the exhaled breath.

THE RUMEN AND ITS MICRO-ORGANISMS

As the utilisation of forages by ruminants depends on microbial fermentative digestion, the principles of digestion in the rumen are discussed as a framework to view the requirements for biotechnology innovations in nutrition. The rumen is the dominant feature of the digestive tract of cattle. This maintains a medium that supports a dense and varied population of microorganisms. These organisms ferment feed materials to produce mainly shortchain organic acids or volatile fatty acids (VFAs), methane and carbon dioxide and the process provides substrate (the feed) and ATP (energy) for the growth of micro-organisms. The microbial mix in the rumen is complex and highly dependent on diet. The main agents that break down fibre, sugars, starches and proteins in the rumen are all anaerobic and include bacteria, protozoa and fungi. The bacteria are the principal organisms that ferment plant cell-wall carbohydrates but the anaerobic phycomycetous fungi may at times be extremely important.

Protozoa are now recognised as having an overall negative effect in the rumen, particularly where ruminants are fed forage diets low in true-protein. Protozoa ingest and digest bacteria and reduce the bacterial biomass in the rumen and consequently the protein supply to the animal. Thus, they decrease the protein to energy ratio in the nutrients absorbed (see later) and increase the requirement of animals for true protein. The net result of the presence of protozoa is an increased requirement for dietary bypass protein and on low protein diets a decreased efficiency of utilisation of feed for growth and milk production. The presence of protozoa in the rumen may also reduce the rate at which bacteria colonise and degrade the ingested feed particles. In studies with sheep fed straw based diets, it has been found that the apparent digestibility of dry matter was increased by 18 per cent after protozoa had been removed from the rumen (*i.e.* defaunated). This research indicates that large increases in productivity may be achieved with ruminants fed fibrous diets, particularly those low in true protein by controlling or removing protozoa from the rumen. Other workers have not seen the differences in digestibility and in some instances removal of protozoa from the rumen has led to decreased digestibility of mixed, starch containing diets.

FERMENTATIVE EFFICIENCY IN THE RUMEN

A deficiency of a nutrient needed by rumen micro-organisms reduces microbial growth efficiency which reduces microbial biomass and eventually reduces digestibility and feed intake, particularly of fibrous feeds. The first priority in feeding ruminants is to ensure no deficiencies in the diet of nutrients for microbial growth in the rumen. Of major importance is that the efficiency of microbial growth (that is, the amount of microbial biomass available for digestion in the intestines per unit of digestible carbohydrate entering the rumen) also determines the proportion of digested feed that is converted to methane and VFA. Methane production accompanies the formation of acetate or butyrate, whereas methane and VFA production are inversely related to microbial cell production.

MEETING THE REQUIREMENTS FOR EFFICIENT MICROBIAL GROWTH IN THE RUMEN

On most diets based on crop residues and low-digestibility forages, the primary limitation to the growth of rumen micro-organisms is probably the concentration of ammonia in rumen fluid The second consideration is deficiencies of minerals, particularly sulphur, phosphorus, magnesium and certain trace minerals. Ammonia in the rumen must be above a critical level for a considerable period of the day to ensure a high rate of microbial growth and digestion and therefore feed intake. The level of ammonia that supports the optimal population of micro-organisms in the rumen the highest protein to energy ration in the nutrients absorbed, and therefore maximum digestion, will vary among diets. In general on forage based diets the ammonia level should be above 200 mg nitrogen/litre. It must be stressed, however, that any nutrient, (including many minerals required in the growth of micro-organisms), that is deficient in a diet will result in low microbial cell yield relative to VFA and lead to a low protein (from microbes) to energy (from VFA) in the nutrients absorbed. The ratio of protein digested and absorbed from the intestines to the VFA produced in and absorbed from the rumen is termed the P/E ratio.

CONSEQUENCES OF THE RUMINANT MODE OF DIGESTION

One of the consequences of the ruminant mode of digestion is that fermentation results in up to 20 per cent of the digestible energy intake being lost as heat and methane. A second major disadvantage is that proteins that are fermented in the rumen are not then sources of amino acids for the animal because they are hydrolysed and their constituent amino acids deaminated by microbes. In general, where ruminants are fed forage based diets typical of that available in tropical developing countries, small amounts of extra nutrients are needed to balance nutrient availability to requirements. Proteins which are directly available to the animals and are protected from degradation increase the efficiency of anabolism of the absorbed nutrients in growth, pregnancy, lactation or work.

QUANTITATIVE ASPECTS OF FERMENTATIVE DIGESTION IN THE RUMEN

The end products of rumen fermentative digestion are governed by the feed, the rate of consumption of feed, the balance of nutrients in the feed for microbial growth and the balance of micro-organisms that develop in the rumen (bacteria, protozoa and fungi). In general, a proportion of the digestible feed dry matter is converted to VFA, methane and carbon dioxide and the balance is assimilated into microbial cells. The pathways of these reactions are well known and a schematic outline is shown in Figure. Microbial cells, that are synthesised from the feed resource use the ATP that is generated in the formation of VFA from the feed to provide the energy for synthesis. The microbes are lost from the rumen pool either by passage out of the rumen to be partially digested in the intestine or by death and breakdown within the rumen (with formation of VFA, CO_2 and methane). Lysis and degradation in the rumen is inefficient as it makes the protein of microbes unavailable as such to the animal. Because microbial cells are more reduced than the substrate fermented, the quantity of microbial cells leaving the rumen per unit of carbohydrate consumed is related to methane production. The efficiency of microbial growth is then a primary determinate of the quantity of methane produced.

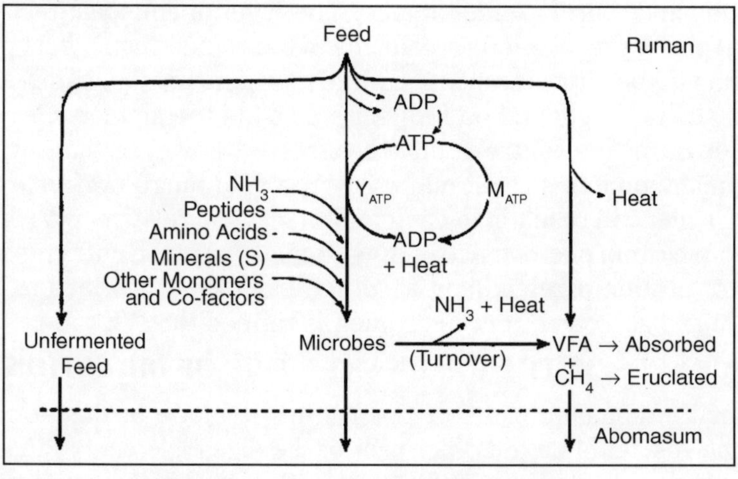

Fig. Energetics of Rumen Fermentation

A Model of Fermentation in the Rumen

For the purposes of the present discussion and to demonstrate the underlying principles of the concepts developed, a model for a 200 kg steer will be used to illustrate the quantitative availability of nutrients from rumen fermentation. The steer consumes 4 kg which represents 25 Mole anhydroglucose or organic matter which is completely fermented in the rumen.

It is assumed:

- That the fermentation of 1 mole of carbohydrate gives rise to either 2 mole acetate, 2 mole of propionate or 1 mole of butyrate, according to the following stoichiometry:

Hexose	→	Pyruvate + 2ATP + H_2
2Pyruvate + $2H_2O$	→	2HAc + $2CO_2$ + 2H2 + 2ATP
2Pyruvate + $8H_2O$	→	2HPro + $2H_2O$ + 2ATP
2Pyruvate + $4H_2O$	→	H Bu + $2H_2$ + $2CO_2$ + 2ATP
CO_2 + H_2	→	CH_4 + $2H_2O$ + 2ATP

- In the stoichiometry, H_2 indicates reduced co-enzymes, HAc is acetic acid, HPro is propionic acid and HBu is butyric acid
- That the animal's rumen is functioning at "normal" level of fermentative efficiency in which one-third of the organic matter fermented is converted to microbial cells and the rest to VFA, CO_2 and CH_4.
- That the moles ATP generated per mole of end-product are for acetate 2, butyrate 3, propionate 3, and methane 1.

On chemical principles, the equation of substrate use and end products from fermentation of 4 kg of carbohydrate is:

- *Carbohydrate to* VFA: 16.7 CHO → 21 HAc + 6HPro + 3HBu + $7.5CH_4$ + 78ATP
- *Carbohydrate to microbial cell precursors*: 8.3CHO → 1.4 polysaccharide + 13.8 pyruvate + $2CH_4$ + 17ATP
- *Overall*: 25CHO → 21HAc + 6HPro + 3HBu + $9.5CH_4$ + 1300 g dry cells.

In the example, one-third of the carbohydrate provides the substrate for microbial cell synthesis 1300 g dry microbial cells are produced at a Y_{ATP} of about 14.5 (Y_{ATP} is a measure of the efficiency of utilisation of ATP generated in fermentation of carbohydrates to VFA; it is defined as the g dry cells produced per mole ATP available.) The upper level of efficiency has a Y_{ATP} of 26. On the other hand the lowest efficiency of a microbial growth in the rumen that is deficient in, say, ammonia, is possibly below a Y_{ATP} of 4. The relative efficiency of the system (indicated as Y_{ATP}) is governed largely by the availability of essential nutrients for microorganisms.

The ranges of Y_{ATP} are shown for:

A. A relatively inefficient rumen (*i.e.* Ammonia deficient)

B. A 'normal' rumen with no deficient nutrient for microbial growth
C. A rumen free of protozoa with no deficient nutrient for microbial growth
D. The theoretical optimum microbial growth efficiency.

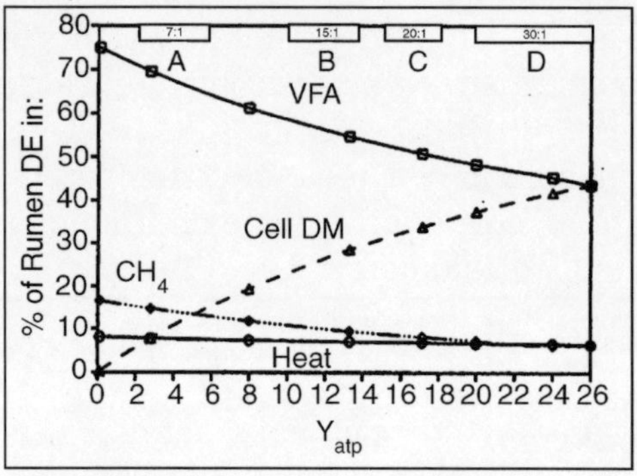

Table. The Effect of Different Efficiencies of Microbial Growth on the Ratio of Protein to VFA Energy (P/E Ratio) Available from the Rumen of a Steer Consuming 4 kg of Organic Matter which is Totally Fermentable.

	Y_{ATP}			
	8	14	19	25
Microbial protein* synthesised (g/d)	500	800	1010	1212
VFA produced (MJ/d)	41	34	30	26
Methane produced (MJ/d)	9.4	8.5	8.0	7.6
Heat (MJ/d)	6.4	5.1	4.3	3.1
P/E ratio (g protein/MJ)	12	25	34	47

* Microbial protein may be only 75–85 per cent digestible and this will change the P/E ratio markedly in the animal.

Based on this model, but assuming a varying efficiency, the microbial cells produced relative to VFA and methane production change are shown in Table. The main point to emphasise is that, depending on the efficiency of utilization of ATP, the amount of carbohydrate converted to microbial cells can be highly variable. It is the efficiency of microbial growth that largely controls the amount of methane produced by an animal.

PROTEIN UTILISATION BY RUMINANTS

Protein that is fermented in the rumen is largely wasted as a source of amino acids to the animal because:

- Dietary protein is degraded and essential amino acids are deaminated to form ammonia and VFA
- Fermentation of 1 g of protein generates only half the ATP that would be produced from 1 g of carbohydrate and therefore anaerobic microbial growth on protein is approximately half that on carbohydrate.

In combination these effects result in only 30 to 60 g of microbial protein becoming available to the animal for digestion for every kilogram of dietary protein that is fermented in the rumen. The fermentation of protein is, however, associated with relatively small amounts of methane production. On the other hand methane is not generated when protein bypasses the rumen.

Protein that is insoluble, or has a high component of disulphide bonds or is associated with tannins tends to bypass rumen fermentation but is digested in the intestines and in this way it augments the microbial protein and alters the ratio of protein to energy (P/E) in the nutrients absorbed. The better the balance of nutrients for microbial growth the higher the ratio of P/E in the nutrients produced in fermentation. The higher the content of bypass protein in the diet the higher the P/E ratio in the absorbed nutrients.

Ensuring a Balanced Nutrition For Ruminants on Forage Based Diets

From the above discussion the first priority for improving the utilisation of a low digestibility forage by ruminants is to optimise the availability of nutrients from fermentative digestion by:

- Ensuring that there are no deficiencies of microbial nutrients in the rumen and therefore the microbes in the rumen grow efficiently and, through fermentative activity, extract the maximum possible amounts of nutrients from the forage (*i.e.* The ratio of microbial cells to vfa produced is high as are production rates of these end products)
- Ensuring that the microbial cells (which provide most of the protein to the animal) synthesised in the rumen are not lysed and fermented in the rumen but are available for digestion and absorption as amino acids from the intestines.

The second objective of a feeding strategy should be to optimise the efficiency of partitioning of absorbed nutrients into product by:

- Supplementation with critical nutrients that escape or bypass rumen fermentation to augment and balance the nutrients absorbed to those required for maintenance of homeostasis, maintenance of body temperature, exercise (or work), and the particular physiological or productive function.

As the nutrients needed for different functions differ in priority, supplementation strategies will need to vary according to climate, environment, management and production targets in any one location.

MAKING A DYNAMIC RUMEN MODEL

Over the past few decades much research has been devoted to quantification of total dietary nutrient needs of dairy cows to support all levels of performance. Many computerized feeding programmes a based on comparison between nutrients delivered in the diet and defined daily nutrient requirements. However, this approach is not sensitive to dynamic changes in feed intake and other factors that impact feed degradability and availability, which ultimately impact the amount of metabolizable nutrients available to the cow to support production.

Nutrient availability from a feed either by microbial fermentation or intestinal digestion is not constant. Competition between inherent feed properties of degradability or digestibility and rate of passage through the digestive tract can explain the amount of a given nutrient pool will be available. This interaction of rates is mathematically described in equations. This concept is the central component of any dynamic rumen model such as the Cornell Net Carbohydrate and Protein System (CNCPS).

- Amount Degraded = Pool size x $K_d / [K_d + K_p]$
- Amount Passed = Pool size x $K_p / [K_d + K_p]$

How much of a given carbohydrate or protein fraction within a feed is degraded will depend upon how rapid the rate of degradation is compared to rate of passage. Rate of passage through the rumen depends upon the fraction. Forages or fibre are slowly passed between 4 and 8 per cent/hr whereas the liquid fraction passes more quickly, between 12 and 20 per cent/hr. If a feed fraction has a rate of degradation that greatly exceeds rate of passage, then most or all of the fraction will be degraded in the rumen. If rate of degradation is slower than rate of passage, then only a small fraction will be degraded in the rumen.

The dynamic components to this system are those CHO or protein fractions where their rate of degradation is nearly equal to the rate of passage. For these fractions, *i.e.*, fermentable fibre CHO and insoluble degradable protein, slight changes in either rate of passage or degradation can greatly influence how much of the fraction will be degraded in the rumen. The following describes a variety of factors that can influence ruminal degradation and passage rates.

Rate of Degradation (K_d)

Degradation of a feed ingredient is inherently related to the chemical and physical properties of all component compounds. Relative to CHO feeds, degradation (ruminal or intestinal) of sugars is extremely rapid compared to starch or fibre. Within starches there is some range of degradation. Fibre is more slowly fermented, while some fibre is not fermented at all. Similarly with proteins, soluble proteins are rapidly degraded while others are more slowly or not at all. It is important then to have an accurate chemical analysis of feed ingredients to determine how much of each of these different fractions comprise a given feed ingredient.

Beyond chemical and physical properties of a feed, degradability can be modified by feed processing, grinding, and rumen conditions. Heating, grinding, ensiling or some combination will increase rumen degradability of feed CHO starch and fibre fractions. For example, corn starch is very crystalline in nature and hydrophobic, thus being fairly resistant to fermentation without processing. If corn grain is ensiled and ground starch availability will be increased dramatically. This explains why coarsely ground corn may be seen in manure compared to ground high moisture corn causing acidosis. Ensiling will increase protein availability to the rumen through increasing solubility. Heat treatment of protein makes the insoluble degradable fraction less rumen degradable. Proteolytic activity in the rumen will dictate extent of rumen protein breakdown of degradable fraction. A decline in rumen pH will greatly reduce fibre fermentation and may reduce proteolytic activity.

Rate of Passage (K_p)

Rate of ruminal passage is calculated for forages and concentrates. Rate of passage will always be slower for forages compared to concentrates. Dry matter intake, forage concentration in the diet and body weight are the primary determinants of rate of passage. Rate of passage increases with increasing dry matter intake and dietary forage concentration. Rate of passage declines slightly with increasing body weight. For each feed an adjustment factor to rate of passage is calculated based on effective NDF of the feed. Passage rate will be reduced with higher eNDF content of feeds.

Putting the Concepts Together

When designing a ruminant feeding programme, concepts of rates need to be addressed in attempting to make sure ruminal availability of energy and nitrogen are coordinated in order to achieve maximal microbial protein production. The goal is to achieve some level of synchrony between ruminal energy and nitrogen sources in order to maximize microbial protein yield. If energy and nitrogen sources are not synchronized, then loss of available energy or nitrogen and reduced microbial activity will occur. Asynchronous energy and nitrogen sources will result in reduced milk production and efficiency and elevated blood urea nitrogen values. In this scenario diseases such as urea toxicity, grain overload or other rumen dysfunctions may occur. If energy and nitrogen sources are synchronized, we maintain maximal microbial protein synthesis. In this scenario not only will the amount of additional dietary protein be reduced, but the cows will have increased dry matter intake and remain healthier, all contributing to increased milk producing efficiency. How is this synchronization achieved?

Intuitively, synchronization of ruminal energy and nitrogen sources is achieved by utilizing various CHO and protein feed ingredients that have differing degradability properties. Use of a single protein (canola meal or soybean meal) or energy source (corn grain) does not give one this flexibility. Using ensiled feeds as the sole forage programme unbalances protein to the more readily degradable fractions. One can alter degrability properties with grinding, heating or some other processing procedure as previously discussed. This is where the strength of the dynamic formulation programmes comes into place. One can use such software to appropriately balance feed ingredients to best meet ruminal needs. In addition, one can use milk or blood urea nitrogen as a diagnostic aid in evaluating the balanced achieved between dietary CHO and nitrogen sources.

Another more simplified way of achieving relative synchrony to the rumen environment is through the feeding programme. Feeding a total mixed ration (TMR) is the single best method of approximating rumen nutrient synchrony. The needed balance between CHO and protein fractions is achieved with multiple meal feeding. This effect accounts for the observed 5-8 per cent improvement in milk production when shifting a conventional to a TMR feeding programme. More slowly degraded CHO fractions can trap readily available nitrogen at a later meal. When concentrates are fed in the milking parlor, this balance is lost. In a conventional feeding system where feed ingredients are fed separately, one then needs to more closely account for differences in CHO and protein availability within feeds. Thus, sequence of feeding can have an impact on milk production response. Forages should be fed prior to concentrate meals to maintain slower rate of passage. Large concentrate meals are best divided and fed 3 to 4 times or more per day to limit the effect of acidosis.

FEEDING RECOMMENDATIONS FOR DIETARY CARBOHYDRATES AND PROTEIN

The preceding discussion has focused on understanding the rumen system and presented concepts on how to better feed the microbial populations to improve milk production efficiency. To move into this new era of dairy cattle feeding, more comprehensive methods of feed composition analysis are needed. With a better understanding of feed composition, we can apply concepts of rates of degradation and passage to a dynamic rumen system to achieve improved feeding practices that will not only improve milk production efficiency, but decrease wasted dietary nitrogen from entering the environment. The following tables provide some practical feeding guidelines in applying these concepts to ration formulation.

NDF and Dry Matter Intake

As previously described, NDF has been shown to influence dry matter intake in ruminant animals. Mertens had shown optimum intake of NDF as a per cent of body weight (% BW) was 1.2 " 0.1 %BW. Neutral detergent fibre intake capacity is influenced by physiologic state and age. Cows can consume more or less dietary NDF, however, there will be negative consequences on production or performance as a result of reduced intake.

Non-fibre Carbohydrates

The NFC portion of the diet is primarily responsible for rapid ruminal fermentation in support of milk production. Milk production will be severely compromised in the face of low dietary NFC content. Conversely, excess dietary NFC content may result in ruminal dysfunction, namely subclinical or clinical acidosis. The balancing act with NFC is to provide sufficient amounts to support maximal microbial growth, yet not providing excess amounts that will reduce ruminal pH and compromise fermentation activity. Table presents suggested dietary concentrations for total NFC and NSC (sugar and starch) components over differing physiologic states.

Protein Fractions

From a dietary protein formulation perspective, nitrogen needs of the rumen should be met first, then metabolizable protein requirements for the cow should be addressed. Dietary crude protein is a poor method by which to formulate diets in this manner. However, one needs some sophisticated dynamic models to adequately predict dietary metabolizable protein delivery.

If one formulates diets to the specific protein fractions, the amount of metabolizable protein should be reasonably adequate. Again, one can only truly predict the degradability and undegradablility of protein sources with a dynamic modeling system, which accounts for dry matter intake and ingredient interactions on availability and contribution to either microbial protein yield or digestible undegraded protein.

Table. Suggested Dietary Concentrations for Various Protein Fractions in Diets for Dairy Cattle.

	Crude Protein	Soluble CP	RDP[1]	RUP[1]
Dry Cows	% DM	% of CP	% of CP	% of CP
Far Off	12 - 13	35	70 - 72	28 - 30
Close-up	14 - 15	30	63 - 65	35 - 37
Lactating Cows				
Fresh	19	30	60 - 62	38 - 40
22-80 DIM	18	31	64	36
81-200 DIM	16	32	66	34
>200 DIM	14	34	68	32

Abbreviations: RDP = rumen degradable protein; RUP = rumen undegradable protein.

In formulating for dietary protein, rumen nitrogen needs must be considered first. Ruminal degrable protein should be between 10.5 and 11 per cent of dry matter to support microbial nitrogen needs. One can never separate rumen nitrogen from CHO needs. To address this interaction between dietary CHO and protein in the rumen, one should formulate for a NFC:RDP ratio between 3.0 - 3.5 to 1.

This suggests as NFC content goes up, then more degradable protein can be assimilated. Rumen protein needs should be met with forage and plant source proteins. If additional bypass protein is necessary, then heat-treated plant proteins or animal proteins (if allowable) can be used to meet the cow's metabolizable protein needs. Use of rumen protected amino acids should be left to only fine tune a diet if necessary, given their cost.

DEVELOPING DETOXIFICATION MECHANISMS IN RUMEN ORGANISMS

Research into the possibility of detoxification of secondary plant compounds was given impetus by the discovery that organisms in the rumen of goats in Hawaii that was able to degrade the breakdown product of mimosine (*i.e.* dihydroxypyridine) in *Leucaena leucocephala,* but ruminants in Australia were unable to do so as they had lost the specific rumen microbe with the necessary enzymes. The microbe isolated from the rumen of the Hawaiian goats degraded dihydroxy-pyridine, the toxic breakdown product of mimosine. Introduction of the microbe from goats in Hawaii into ruminants in Australia prevented the symptoms of mimosine toxicity in previously susceptible ruminant livestock. This has raised the possibility of genetically engineering into rumen organisms the ability to degrade such toxic compounds. Research at the University of New England is in progress to attempt to introduce genes into bacteria for defluorination of fluoroacetate, a lethal compound present in *Acacia gidyea,* which often kills cattle in northern Australia when consumed in the dry season. The successful transformation of rumen bacteria with a single gene for defluorination of fluoroacetate would indicate the possibility of the utilization of a range of other plants, which may be weeds (*e.g.* lantana), or potential high biomass producers (some legumes and legume trees) and also high protein crops (*e.g.* canavalia).

The problems are universally the same. The first is identification of bacterial DNA encoding for the suitable enzyme followed by development of methods to introduce this into an appropriate bacteria, stabilisation of the DNA in the cell and the last problem is whether the modified organisms will survive in the rumen and be capable of expressing the gene. The major research area would be in finding, isolating and packaging the DNA for the appropriate enzymes which must be capable of expression of being switched on easily and function in an organisms in the rumen ecosystem.

There is a danger in developing the ability of ruminants to degrade toxic compounds, in that often these are the last defence of plants being eaten and in the arid areas could lead to deforestation and soil erosion.

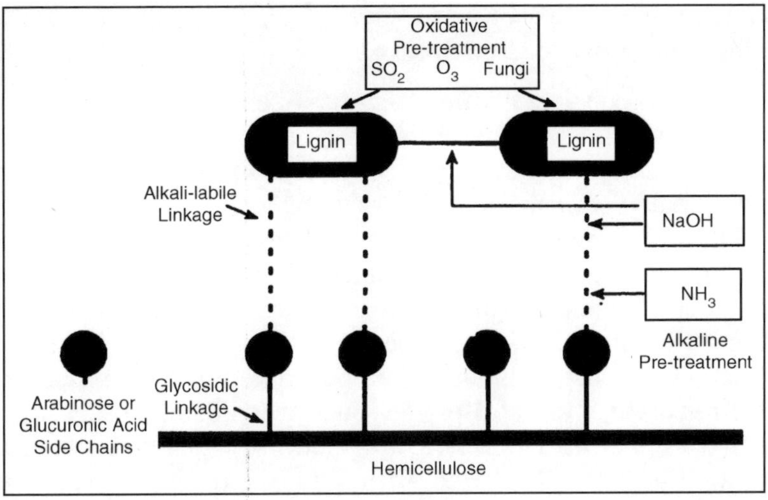

Fig. Sites of Action of Chemicals that, when Added to Straw, Improve its Digestibility

TREATMENT TO INCREASE DIGESTIBILITY

Improving the Digestibility of Crop Residues by Chemical Treatments

There are a number of acid and alkaline hydrolytic processes that will solubilise lignin, cause disruption of fibre by swelling or improve the potential digestibility of fibrous roughages in other ways. It seems that there is adequate information available on the technology of treatment and on the benefits of treatment of low quality forages. There is therefore little stimulus in pursing research of the methods.

A new and novel approach will at times, however, require research but at present it can be safely assumed that alkaline hydrolysis of wet straw by ammonia generated by urea (usually 4 per cent of the dry matter) by microbes in the straw, is probably the most feasible treatment for application in developing countries. The use of anhydrous ammonia application is also potentially applicable.

The ammonia in the straw has two effects:

1. It improves the fermentability of the straw in the rumen by breaking some of the ligno-hemicellulose bonds and
2. It provides non-protein-nitrogen for growth of microbes in the rumen when the forage is consumed.

Although urea ensiling of straw can be economic and effective the technology has not been widely taken up by small farmers in developing countries in general, although there are a few reports of individual farmers regularly using the technology. It is necessary for researchers to understand the reasons for non-acceptance of such technology which is almost always associated with a slow return on funds expended and sociologic constraints.

Microbial Treatment of Straw and Other Crop Residues to Improve Digestibility

In recent years considerable emphasis has been placed on the improvement of fibrous crops by the strategy of growing non-toxic fungi on the straw. In particular, the white rot fungi have been used because of their ability to delignify the plant material. The major problem that has arisen is that the technology, although relatively simple, is more complex than that of urea treatment of roughages and although protein is produced and digestibility of the residue increased, a considerable proportion of the total biomass is lost, particularly when white rot fungi is allowed to proceed to the mushroom formation stage. To some extent the loss of biomass can be prevented by a reduction in the incubation time of the straw and forages, such that biomass losses are reduced to less than 10 per cent. Dr. B.N. Gupta and his colleagues at the National Dairy Research Institute (NDRI) in Karnal, India have developed a two stage process using an altophilic fungus (*Coprinus fimetorius*).

The technology is two-stage and relies on:

1. Alkali treatment (urea ensiling)
2. A second incubation period where a fungi is cultivated on the straw.

The technology is relatively simple, although the two stages make it more complex than the urea treatment alone for application by the small farmers. Both chemical and microbiological treatments may, however, be very effective where large amounts of feed must be stored and moved to areas where forages are scarce because of drought. However, again caution is needed in that to obtain maximum benefits from such feed, cattle will require supplements of bypass protein.

Potential for Increasing the Digestibility of Poor Quality Roughage by Manipulating the Digestive Physiology of the Animal

Goats compared with sheep, and buffalo compared to cattle, tend to retain fibrous materials in the rumen for longer periods and extract a few more units of digestibility at the same time. Within these comparisons, goats and buffaloes recycle a higher amount of urea-N back to the rumen. In the same way, *Bos indicus* cattle recycle more urea-N back to the rumen that *Bos taurus* cattle. Mechanisms that slow rumen turnover, increase rumen volume and allow greater recycling of urea-N may therefore improve the utilisation of poor quality forages.

Immunisation against somatotropin release inhibiting factor (SRIF) has recently been demonstrated to increase flow rate through the gut, leading to increased digestibility and greater availability of protein to sheep on concentrate/forage diets. The effects of immunisation against SRIF on cattle on forage/bypass protein diets would be extremely interesting. However, the technology will not be easily applied to village farmers because of the numbers of injections required at the present time to effect immunity.

ALTERING THE PARTITIONING OF NUTRIENTS WITHIN THE ANIMAL

A whole range of injectable compounds will lead to more efficient utilisation of nutrients by livestock. In particular, repartitioning of nutrients into lean meat production or into milk production. These include anabolic steroids, growth hormone, and other hormonal growth factors, beta agonists and techniques such as the suppression of hormones by immunisation. These compounds in general, act through manipulation of hormone status and they lead to improved lean meat in the carcass and a greater efficiency of feed conversion into product. For a recent review of this area, the reader is referred to the publication by Buttery and Dawson.

Anabolic Steroids

Natural sex steroids are used to promote growth and reduce the fat/protein ratio in meat in many countries of the world but their use has been recently banned in Europe. The injected hormone alters the influences of endogenously secreted hormones (oestradiol injections into cattle usually lift growth hormone levels in blood) and growth hormone is more tightly bound to sites in the liver. As is usually the case there is a marked interaction with nutrition and on imbalanced diets in general the effect of exogenous hormones is less obvious than for instance when a diet of silage is balanced by a bypass protein (fish meal). Again this stresses the differences and therefore the need for a new approach to biotechnology in developing countries where ruminant diets are usually low in true protein.

Growth Hormone

Recent developments in fermentation technology together with recombinant DNA technologies have made it possible to produce large quantities of peptide proteins which appear to be the same as bovine or porcine growth hormone (BSt. or PSt.). These growth hormone injected into animals stimulates growth and dramatically decreases fat content of the animals carcass. They also stimulate milk production by some 20–25 per cent in dairy animals. In the initial studies with growth hormone injections milk production increases appeared to result from repartitioning nutrients and mobilisation of nutrients from tissue deposits in recent times a long term feed-intake effect has been reported. In addition the stimulation of milk production was dependent on the level of management and was not apparent in herds with poor management. In Italy and India in well managed buffaloes, milk yield has been increased by exogenous growth hormone by up to 25 per cent but the cost of the injections may limit this application.

β-Agonists

β-agonists given by mouth, increase lean meat deposition and reduce fat content of the carcass in pigs and ruminants but with only small increases in liveweight gain. Again it is more than likely that the responses are dependent on the nutrition of the animal.

Injected Growth Promotants Versus Supplementation to Balance Nutrition

There are strong similarities between the response observed to growth hormone/β-agonists etc. and the response to correcting the P/E ratio in nutrients absorbed. At least one paper reports an interaction of bypass protein and response to exogenous hormones. It is thus possible that the optimisation of P/E ratio in the nutrients absorbed elicits responses in hormonal release that mimic the injection of growth hormone etc. Conversely and if true, this means that effects of exogenous hormones may not be so obvious where the P/E ratio in nutrients absorbed is optimised. In addition the requirements for protein relative to energy in the transgenic animals expressing growth hormone will be much increased as will requirements for minerals and vitamins.

The important conclusion from the paper by Buttery and Dawson is that: "the current technology has given many exciting possibilities but there are, however, still major gaps in the knowledge of the major metabolic control processes associated with growth which will have to be filled before a universally accepted method of growth promotion is developed"

This is a reservation which is repeatedly used by biotechnologists and indicates that there will be an ongoing requirement for basic research at the level of metabolic control.

OPTIMISING MICROBIAL GROWTH IN THE RUMEN

MINERAL REQUIREMENTS OF RUMEN MICROBES

The rumen microbes have specific requirements for both macro and micro minerals to meet the needs of structural components of cells and for components of enzymes and co-factors. Little is known about the requirements of the microbial milieu for trace elements and as a 'rule of thumb' it is accepted that if the animal is not deficient then it is unlikely that the rumen microbes will be deficient.

As Suttle has so aptly put the situation, "it will rarely be possible to approach a suspected mineral deficiency situation with a table of minor nutrient requirements or biochemical criteria in the hand, and define a scale of the animal health (microbial health) problems". In practice, either no mineral supplements are used or a "shot gun" mixture is given as a salt licks or as molasses (which is concentrated plant juice rich in minerals) suitably fortified with minerals. As with any deficiency of a nutrient the likely scenario of a mineral deficiency for rumen organisms is first a reduced growth efficiency of microbes (lowered ratio of cells to VFA produced) with or without a decrease in digestibility. As the deficiencies become more extreme the digestibility of forage must decrease along with the decrease in microbial pool size and it is only then that feed intake will decrease. Feed intake however, will be decreased as P/E ratio decreases if the animal is heat stressed (see later). Correction of deficiency will obviously reverse these effects.

Requirements for Ammonia

Ensuring adequate ammonia N in the rumen to supply the majority of nitrogen for microbial growth is the first priority in optimising fermentative digestion of forage. Satter and Slyter suggested that 50–80 mg NH_3-N/1 of rumen fluid was the optimum for maximising microbial growth yield and this has been widely accepted. However, recent studies from two laboratories in Australia have clearly indicated that the minimum level of ruminal fluid ammonia for optimum voluntary intake of low N, low digestibility forage by cattle is about 200 mg N/1, even though the digestibility of the forage (in nylon bags) was optimised below 100 mg NH_3-N/1. All these studies were carried out in hot environments and the effects on feed intake are possibly explained by an improving P/E ratio in the nutrients absorbed, which reduces the metabolic heat load.

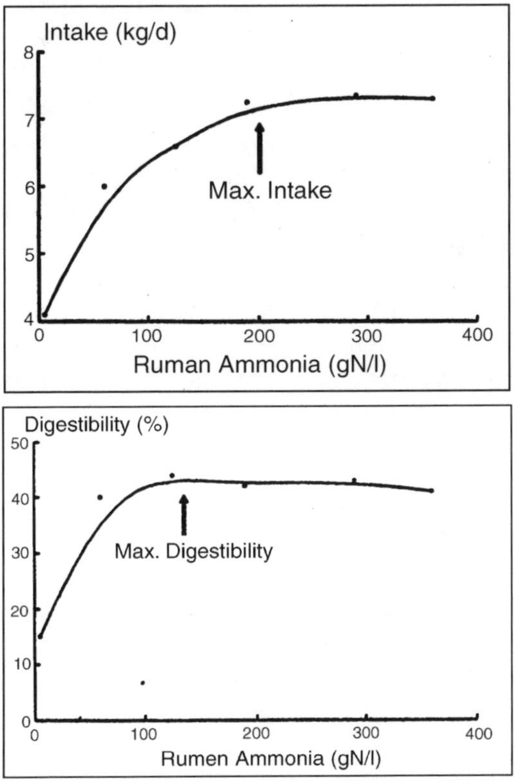

Fig. The Effects of the Level of Ammonia in the Rumen on the Intake and *in Sacco* Digestibility of Straw by Cattle. The Ammonia Levels were Adjusted by Infusing Urea in the Rumen.

The effects of increasing ruminal fluid ammonia by infusion of urea into the rumen of steers on the intake of rice straw and its digestibility in nylon bags in the rumen are shown in Figure.

Timing of Urea Supplements and the Ratio of Sugars and Starches to Fibre in a Diet

Supplements must provide adequate levels of ammonia in the rumen for continuous growth of both fibrolytic and saccharolytic organisms. The only satisfactory approach to meeting these changing requirements for ammonia is to provide ammonia continuously. One way of doing this is to provide salt/urea or molasses/urea licks and allow the animal to take these as needed. There are indications that cattle and buffaloes given continuous access to multi-nutrient blocks based on molasses/urea are able to control fairly closely their intake of urea. Once buffaloes were accustomed to molasses/urea blocks they adjusted their intake according to the N content of the basal diet (NDDB—unpublished data). Lambs given wheat straw and molasses/urea blocks also had similar abilities and consistently maintained their rumen ammonia levels above 200 mg N/1.

Requirements for Amino Acids/Peptides by Rumen Organisms

There has been considerable controversy concerning the requirements for peptides/amino acids by rumen microbes for efficient growth. A number of studies, however, have reported results of *in vivo* studies which appear to have indicated no apparent requirement for amino acids for efficient growth of rumen organisms. The results of studies by Maeng *et al.* may explain some of the contradictory results. The studies of these researchers indicated that rumen microbes growing on different carbohydrate substrates have requirements for different N-substrates; celluloytic organisms may not require amino acids to the same extent as organisms growing on starches or sugar as the major substrate. For microbes utilising sugars growing on starches there was an apparently high requirement for preformed amino acids/peptides but this was not so for cellulolytic organisms. Maeng *et al.* also showed an increase in efficiency of microbial growth on fibrous carbohydrates with decreasing dilution rate of rumen contents. If true this may be advantageous to ruminants given low quality forages that must be retained in the rumen for a considerable period to allow digestibility to be optimised. At the same time the improved ratio of cells: VFA yielded (*i.e.* P/E ratio) along with the increased availability of the total nutrients are both advantageous.

Such a mechanism would advantage an animal with:

- A comparatively slow turnover rate of rumen contents (*i.e.* Buffalo vs. Cow or goat vs. Sheep
- Fauna-free vs. faunated animal or animals at high environmental conditions versus cold stressed animals.

Amino Acid Requirements of Microbes Digesting Fibre

The organisms in the rumen that are largely responsible for the fermentation of cellulose (*Ruminococcus albus, Ruminococcus flavefaciens* and *Fibrobacter succinogenes* (previously called *Bacteroides succinogenes*) appear to have minimal requirements for amino acids and grow on ammonia. Conversely, organisms important in starch hydrolysis (*Butyrivibrio fibrisolvens, Bacteroides ruminicola, Selenomonas ruminantium, Streptococcus bovis* and *Ruminobacter (Bacteroides) amylophilus* readily incorporate amino acid N and in many cases peptides.

Supplementation of sheep fed a poor quality forage with branched chain VFA has been reported to increase the apparent flow of microbial-N to the duodenum. The apparent stimulation of microbial growth with branched chain VFA has also been shown to increase feed intake on occasions. This together with the suggested requirements for peptides/amino acids by rumen organisms (which on the basis of the results of Maeng *et al.* must now be questioned) has tempted many scientists to explain the increased feed intake of ruminants, on poor quality forages that are supplemented with bypass protein, to the slow release of amino acids, peptides and branched chain fatty acids to the rumen milieu from the protected protein, even though in most studies there was no evidence of increased digestibility with such supplements in predominately fibre based diets.

The above discussion indicates that the cellulolytic organisms in the rumen even of cattle on straw based diets, are rarely if ever deficient in amino acids, peptides or branched chain VFA in the rumen. This is not to deny that these organisms may still need these nutrients in "catalytic" amounts but that they are rarely if ever at such low concentrations in rumen fluid to bring about a deficiency.

The Roles of Small Amounts of Fresh Forage in Straw Based Diets

Farmers in developing countries have generally recognised the benefits to cattle of adding a small amount of fresh green herbage to straw based diets. These practices, which have evolved through trial and error, may have a number of beneficial effects which include the supply of vitamin A, essential minerals, ammonia, peptides/amino acids in an otherwise unsupplemented diet.

Recently it has been shown that where the supplemental forage in a straw based diet given to sheep is of high digestibility a boost to digestibility of the basal diet occurs even at relatively small levels of supplementation. The rate of digestibility of straw depends on the rate and extent of colonisation of fibre and the biomass of adherent organisms and the high digestibility forage supplement may act to seed microbes onto the less digestible straw. On the other hand, other influences cannot be ruled out. For example, in the studies of Silva and Ørskov in the absence of an effect of supplemental forage on digestibility, the rumen ammonia levels were often not significantly below 200 mg NH_3-N/1. Where increases in digestibility of the basal forage occurred to supplemental forage the ammonia levels in the rumen were significantly below 200 mg N/1 and the supplement apparently improved the concentration to above the critical level.

Elimination of Rumen Protozoa and Preservation of the Fauna-Free State

The physiological research which has shown that there is an increased availability of microbial protein for digestion in fauna-free as compared to faunated ruminant has supported the feeding trials with large numbers of animals which has demonstrated significantly high feed conversion efficiencies and wool growth rate in sheep without rumen protozoa as compared to control animals.

Fauna free cattle on the same intake of a low protein, molasses/urea based diet grew at a 43 per cent greater rate than faunated cattle on the same feed intake. The improved production was therefore an effect of a higher efficiency of feed utilisation. The discussion to follow, on the implications of environmental temperature/humidity for the nutrition of ruminants, will indicate why a major change in P/E ratio in the nutrients absorbed (*i.e.* the major difference between faunated and fauna-free ruminants) will be more effective in improving ruminant production in the tropics as compared to temperate/cold countries. In the tropical areas the available forages used to feed to ruminants are generally lower in digestibility, lower in true protein and animals are rarely cold, but heat stress at times may be severe. It should also be noted that animals brought into animals houses from fairly could environments may at times suffer severe heat stress through a combination of a well insulated fleece or coat and an imbalanced diet.

MORINGA FODDER IN RUMINANT NUTRITION

Within the framework of market-oriented system improvement programme research at the International Trypanotolerance Centre, Banjul, the need to improve the feed resource base as an ameliorative move across the farming systems is overwhelming. This is principally a response to one of the most crucial needs of the livestock sub-sector – nutrition. From time immemorial, the groundnut residue has been the traditional feed resource of choice especially in urban areas. These crop residues are hauled into peri-urban animal depots where they are used on a zero-grazing basis. However, due to economic and agronomic reasons, during the last decade, it has become necessary to investigate the potentials of alternative feed resources. Consequently, the International Trypanotolerance Centre, Banjul screened a cohort of more than thirty five feed resources that can be used by urban ruminants as basal and/or supplement diets.

The Moringa oleifera variously known as drumstick tree, etc has been systematically investigated during the last three years at ITC. The *Moringa* plant is well known for its enormous biomass production and it promises to be the plant of the future in ruminant animal supplementation strategy. Although not completely strange in the West African biosphere, this grossly underexploited plant has a lot to offer as a food and fodder resource in the sub-region. Considering the agro-ecological characteristics of the sub-region in general and The Gambia in particular, the plant needs to be further investigated and modalities of integration into the farming system carefully conducted on the merits of the respective locations.

The plant *Moringa oleifera* is very popular (locally known as 'never die') in the Senegambia region where it exists principally in scattered uncultivated forms. It is mainly consumed in various forms as food in the region although a few useful extra-culinary attributes have been suggested. Reports from other tropical countries around the world, especially in Latin America and the pilot trials conducted in The Gambia (Adediran and Akinbamijo, unpublished) have shown very encouraging biomass yields. More recently, under high density cultivation, biomass yields in excess of 15 tonnes DM/ha in a 60-day growing cycle has been obtained at the International Trypanotolerance Centre, Banjul. This volume of high quality biomass is simply overwhelming given the context of the semi arid ecologies in The Gambia. This is of particular interest to animal nutrition where dietary protein sources are becoming increasingly expensive and difficult to access.

Given this magnitude of potentially useful biomass, there has been no systematic attempt to exploit *Moringa* either in terms of its agronomic attributes or its nutritional values. The preliminary investigations were therefore aimed at the systematic investigation of the nutritive evaluation and the modalities of establishing *Moringa oleifera* in the various farming systems in West Africa From the growing body of information on the plant, it is common knowledge that it is endowed with exceptional physicochemical properties with the possibilities of high biomass yield described above obtained at ITC.

In addition, it is known to be devoid of any known anti nutritive factors and insignificant tannin contents. The plant is known contain at least 25 per cent protein. Compared with other conventional ruminant feedstuffs in The Gambia, it has a very high biological value and considerable potential for adoption as food for humans as well as ruminant fodder resource

SYSTEMATIC INVESTIGATIONS IN THE GAMBIA

Since the beginning of the century, a new research strategy on the *Moringa* initiative was launched in The Gambia. This initiative received a lot of logistical support from the Institute for Animal Production in the Tropics and subtropics of the University of Hohenheim and the Church World Service, Dakar and the

National Nutrition Agency of The Gambia. With the compelling attributes demonstrated in The Gambia, there was every reason to pursue the *Moringa* research plan in the context of the Senegambian faming system. Within this system, *Moringa* will readily fill both human and animal nutrition gaps. The ultimate beneficiaries therefore appear to be poor agro-pastoralists that need enormous food and feed support to keep their lives and those of their animals (upon which their livelihoods depend) in a sustainable manner. The nutritive value of *Moringa oleifera* is currently undergoing vigourous tests on-station and on-farm in parallel in the context of the West African farming systems and ecology.

AGRONOMY

The *Moringa* plant is cultivated in a high density mode with three approaches adopted in the evaluation of *Moringa oleifera* in the West African farming system context. The first step was the agronomic assessment of biomass yield under diverse production and harvesting regimes. Following the encouraging results obtained from a pilot study on *Moringa oleifera* establishment and yield at the International Trypanotolerance Centre, Banjul in 2001 with a planting density of 15cmX15cm, a logical next step was the determination of the optimum cutting height and intervals that would provide the best quality of fodder as well as retain the biomass yield at a relatively high level. In theory, we were to determine the maximum biomass yield without significant negative impact on the re-growth potential of the plant under reasonable conditions that can support re-growth. Three cutting heights – 20, 30 and 45 cm were tested in three cutting interval cycles of 50, 100 and 150 days after planting. Fodder was assessed for biomass yields, fibres and digestible nutrients.

Biomass Evaluation

Biomass yield obtained by estimate dry matter content of the materials recovered in a 1 sq meter quadrat indicated the possibility of obtaining up to 20 tonnes DM/ha in a 50 day growing cycle. However, for the purposes of easy communication to the ultimate beneficiaries, it is recommended that the cutting cycle be rounded to once in two months rather than every fifty days. With a cutting interval of only 60 days before it is used as fodder, the nutritive value of the fodder is very high with very low chances of lignification. It must be stated however that optimum growing conditions such as continuous irrigation, high planting density and a fertilizer regime of 50 kg/ha/month are imperative before such high biomass yields can be obtained.

In vitro Assays

In order to improve our understanding of the nutritive value of *Moringa*, it was compared with other *in vitro*-simulated combination diets in a series of laboratory-based studies using the Hohenheim Gas Test method to quantify corresponding rumen gas production to *Moringa*-based diets. Emerging trends indicate that the *Moringa* option compared with groundnut cake, has a significantly higher gas production potential of up to 36 ml gas compared with 28 ml of VFAs at 40 per cent inclusion in the diet. Using the Hohenheim Gat Test procedure, we have been able to improve our understanding on the comparative advantages of *Moringa* on groundnut hay.

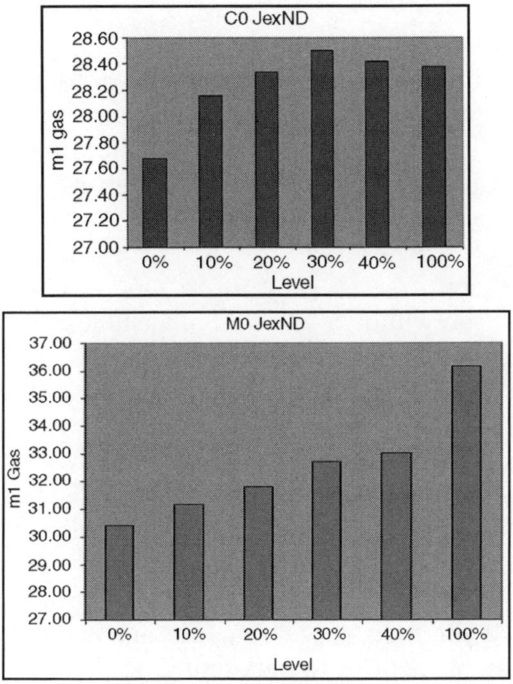

Fig. Gas Production

Ruminant Nutrition: High density nutrient supplementation strategy

Moringa oleifera was investigated as an alternative to conventional concentrates in defining an appropriate supplementation strategy in a dairy production system using two classical methods:

- *In vivo* digestibility
- *In vivo* validation trial

In addition, animal response on a trial diet using *Moringa* to supplement a groundnut hay- based diet has shown that there is no significant difference in performance when compared with animals supplemented with groundnut cake based concentrate.

In a short term animal response study set up to validate the *in vitro* results using growing crossbred animals, in quantitative terms, animals offered *Moringa* leaves as supplements did attain a higher growth rate of 440 g/d compared to their counterparts offered conventional locally available concentrates (1:1 mixture of GNC and rice bran) that grew at the rate of 385 g/d although this difference did not attain statistical significance (Fig 2). As expected, the supplemented animals offered either *Moringa* or concentrate out-performed their counterparts offered only groundnut hay and gained only 274 g/d. The growth rate of the control animals was significantly lower than those of their supplemented counterparts.

LESSONS LEARNT

The present results confirms the enormous potentials of *Moringa* as a supplement for ruminants especially in highly intensive production systems such as the peri-urban dairy systems and as ameliorative dry season feeding strategy in the traditional extensive system of husbandry.

Challenges

- Processing of biomass to avoid post harvest losses especially during the rainy season

On-farm production will be a seasonal affair unless for exceptional cases where there are farmers that can afford to irrigate their *Moringa* fields. Our experiences in The Gambia with on-farm intensive production are that there is a high likelihood of excessive biomass yield during the rainy season. In order to optimize the exploitation of *Moringa*, the issue of fodder preservation becomes imperative in order to avoid post harvest losses. Efforts on how to address the perishability of the fodder is a current subject of research at the International Trypanotolerance Centre, Banjul.

- *Storage for use during the dry season:* As on-farm production is geared towards food and feed, appropriate storage and methods and facilities have to be in place prior to harvesting. It is common knowledge that *Moringa* is a nutrient-rich supplement in terms of minerals and vitamins essential for vital physiological processes in humans and livestock. In order to avoid post harvest depreciation in the nutritive values of the nitrogen free extracts, there is need to provide adequate long-term storage facilities available to *Moringa* growers.
- *Year round production in intensive systems:* In peri-urban dairy systems, where feeding is required to cater for the different phases of the dam cycle, appropriate cow nutrition becomes necessary.
- Land availability and communal ownership of high density *Moringa* fields

The introduction of high density *Moringa* cultivation is still a novel approach to most of the end users. Although the plant itself is not new, the end users have never needed to have dedicated plots for growing *Moringa* in large quantities. As land tenure remains a big issue in areas where *Moringa* is most needed – especially as periurban feed resource, it is important that policy instruments in urban areas are formulated with feed resource concerns in perspective. In the rural areas where *Moringa* can greatly improve human and animal nutrition on community basis, there is need to adopt community forest ownership strategies to enhance the adoption and establishment of communal *Moringa* fields.

EFFECTS OF TANNINS ON RUMEN FUNCTION

It has been believed for some considerable period that tannin above 5 per cent can become a serious anti-nutritional factor in plant materials fed to ruminants (McLeod, 1974). Barry (1983) and his colleagues have demonstrated with *Lotus pedunculatus* that the ideal concentration of condensed tannins in this forage legume is between 2–4 per cent of the diet dry matter, at which level they bind with the dietary proteins during mastication and appear to protect the protein from microbial attack in the rumen. On the other hand, other forages with tannins have anti-nutritional effects in the rumen and reduce N retention (Waghorn and Shelton 1995). If the protein-tannin complex dissociates under acid conditions then the protein can be digested in the lower gut. At higher levels (5– 9 per cent) tannins become highly detrimental (Barry, 1983) as they reduce digestibility of fibre in the rumen (Reed *et al.* 1985) by inhibiting the activity of bacteria (Chesson *et al.* 1982) and anaerobic fungi (Akin and Rigsby, 1985) high levels also lead to reduced intake (Merton and Ehle, 1984); above 9 per cent tannins may become lethal to an animal that has no other feed (Kumar, 1983).

Sheep have been shown to adapt slowly to tannins in a diet of *Acacia* leaves, suggesting that there are rumen organisms that in some way detoxify their effects (Reed *et al.* 1985). Recent studies have demonstrated a bacterium in the rumen of goats and sheep capable of growth at high tannin levels in the rumen (Brooker, J., personal communication). Thus a little tannin has been usually accepted as being able to protect protein of forages and allow a higher efficiency of feed utilization by the animal. However, recent results throw some doubt on this. Tannins may indirectly effect rumen function by reducing rumen ammonia levels through decreased protein degradation in the rumen. If rumen ammonia levels decrease below 80 mg N/l then fibre digestibility may be depressed and digestibility is reduced well below 10 mg N/l (Leng *et al.* 1993). Whenever tannins are present in forages there may be a need to supplement ruminants with a non-protein nitrogen source such as urea or chicken manure. Conversely, tannins in feed may increase detrimental effects on rumen function when the basal diet is low in protein.

Tannin in Plant Foliage

Condensed tannins are present in only some plant species. In general, shrub and tree foliages are likely to be higher in tannins than pasture plants, and leguminous forages from the tropics are generally higher in tannin than those from the temperate countries. The level of tannins within a species has been found to vary considerably depending on a number of factors. For example in New Zealand, *Lotus pedunculatus* grown on fertile, high moisture soils has about one third the condensed tannin content of *Lotus* grown on hill country under water stress (Barry, T., personal communication).

The literature in this area is often confusing as reported levels of tannins often seem to be higher than can be explained and these values often are rejected by reviewers as possibly due to errors of analysis. For example, *Prosopis* leaves have been reported to have tannin levels of 2.2 per cent of the dry matter (Sehgal, 1984) but in 15 individual trees Joshi *et al.* (1985) reported levels ranging from 10.6 to 25.3 per cent which suggested to some authors that there may have been major analytical difficulties. Certainly the level of tannins in the foliage of trees is highly variable and depends on environmental stress (fertility, soil-water relationships, insect attack etc.). Newer information also points to analytical flaws in preparation of leaf material for analysis and drying reduces the measured tannin levels over fresh material. New leaves often have higher tannin content than older leaves (Vaithiyonathan and Singh, 1989) and in South Africa the tannin content of*Acacia* grazed by Kudu increased with grazing pressure (van Hoven, 1991).

Tannins at 14–16 per cent are present in the bark of many trees (Dalziel, 1948) indicating a very large pool in the tree that can possibly be mobilized. In general it could be expected that the green bark of new growth would contain less tannin than the brown bark, and the leaves and petals less than bark.

Tannin Mobilization

There is a small amount of literature indicating that the tannin content of some tree foliages may be controlled in some way and can be elevated at times of high risk of defoliation (*i.e.*, by insect attack, cutting and harvesting, or grazing).

An investigation in South Africa of the death of a number of Kudu on a wild life farm has led to knowledge that may have major implication for management of some fodder trees (van Hoven, 1985). A number of Kudu (member of the deer family) died after grazing on a small area of *Acacia* trees (van Hoven, 1991). Subsequent studies have shown that in the wild these Kudu would approach such woodlands and after grazing on the trees on the periphery move quickly to trees well separated from those recently grazed. However, because of the enclosure these animals were forced to consume more of the foliage from one woodland. Subsequent studies implicated tannin in the death of the animals and this led to a study of tannin in the tree foliage. Tannin levels rose sharply over a 15 minute to 1 hour period, not only in the trees grazed by deer, but in the trees adjacent to those that were being damaged.

Nomads in India, grazing camels on *Prosopis juliflora* on the roadside carefully explained the need to move their camels to fresh trees after a short time of grazing in the one area (Leng, R.A. personal observation). Although not yet tested, this fits the idea of some trees responding to grazing by increasing the content of anti-palatability secondary plant compounds. The implication is that there is a pool of tannin which can be readily mobilized by activation of specific enzymes sensitive to air borne materials released from damaged foliage. This obviously has been an important survival mechanism for trees on the savannahs. Also it has major implications for the use of browse trees in pasture/tree associations for cattle production. However, responses of some tree foliages to such treatment is far from clear and a number of trees do not respond to damage in any way. This area requires experimental work in a number of situations. The effect of simulated grazing on tannin content of a number of tree leaves is shown in Table. The information on such responses in a number of trees needs to be compiled.

The Implications of Tannin Build-up in Foliage

Tannin buildup in foliage has enormous implications for the use of fodder trees. It may partially explain some of the contradictory nature of much of the data on the efficiency of use of fodder trees, and the common rejection of trees for fodder purposes in some areas and not in others. The quality of harvested foliage will always depend on where it is produced, how it is harvested, the stage of plant growth, the climate, the effects of insect damage and the stocking rate. Where trees respond to damage, it is possible that occasional grazing by animals in a plantation may effect tannin levels in the leaves for up to 4 days.

As tannins appear to be both anti-nutritional and perhaps nutritionally beneficial it is important to determine what factors influence tannin levels in foliage; whether these factors can be controlled (*e.g.*, strategic fertilizer or water application) and whether harvesting techniques and stocking rates can be developed to optimize or minimize the tannin levels. If it is too complex to develop such techniques then it is obvious that high tannin levels may be acceptable so long as they are diluted by other feed resources that have sufficient protein to bind free tannins beneficially but this necessitates a cut and carry system for tree fodders.

Tannins in the Rumen

The fate of tannins following ingestion by ruminants depends on the type of tannin. Most tannins form complexes with protein in the plant material, in saliva or the rumen contents. Hydrolyzable tannins hydrolyze in gastric acidity beyond the rumen, releasing protein, amino acids and small units of phenolics that probably pass to the urine. At high levels of tannin intake, both mucoproteins and the epithelial cell lining of the digestive tract are affected.

This alters the integrity of the gut wall causing problems of gastritis, slowed propulsion of feeds and constipation (Kumar and Singh, 1984). Only under exceptional circumstances will herbivores consume large quantities of forages containing tannins, although recent studies from Zimbabwe suggests that the intake of browse can be increased by feeding very small quantities of polyethylene glycol (Duncan, 1994). However, from the point of view of this publication, fodder trees are considered only as potential supplements to poor quality forage and not a basal feed reserve.

Drying promotes the combination of tannin and plant protein before ingestion which may result in different responses by ruminants on forage diets to supplements of dry or fresh tree foliage. Proanthocyanidin tannins are held in special organs in the leaves to prevent their interference with the plant's own metabolic apparatus, and this factor favours tannin-tolerant animal browsers that have tannin-binding proteins in their saliva (Austin *et al.* 1989). In these instances the salivary factor binds the tannins and spares valuable forage protein with a higher concentration of essential amino acids.

Attempts to use tannins to promote rumen escape ("bypass") of feed protein have had limited success. Fresh forage may be ingested before tannin-protein complexes can oxidatively cross-link, as in the case of dry feeds, although drying has been shown to have favourable effects on productivity. Recently, Waghorn and Shelton, (1995) showed that, in sheep fed freshly cut rye grass, the addition of a third of the diet as *Lotus pedunculatus* (which supplied 1.8 per cent condensed tannin in the total diet) resulted in the digestibility of the protein being significantly decreased from 78 per cent to 65 per cent. The inclusion of *Lotus* forage also lowered dry matter digestibility by 3–7 per cent, which was almost all accounted for by the lowered protein digestibility.

Apparently the level of protein availability in the animal from plant and other sources was unaffected, as wool growth and liveweight gain were not different in groups given *Lotus* forage compared with those on pure rye grass. These experiments suggest that condensed tannins overprotect protein and that the precipitation of protein-tannin complexes protect the micro-organisms in the rumen from the detrimental effects of condensed tannins. The presence of tannins in forages stimulate salivary flow in animals. A number of studies have shown an apparent increase in microbial protein leaving the rumen after feeding moderate levels of tannin (Beever and Siddons, 1986), but the level of increase is not convincing, considering the technology used to make the measurements.

Nitrogen balance is apparently improved in animals that are fed low levels of tannins, although digestibility of forage fibre may be lowered. The effect of condensed tannins overall appears to be to make more amino acids available in the intestine. Whether this is the result of increased microbial growth efficiency or increased dietary protein availability is unclear.

Plant Compounds in Fodder Trees

Many foliages have chemicals that appear to be produced for the purpose of deterring invasion or consumption of their leaves by microbes, insects and herbivorous animals. Whilst tannins are the best known of these, there is a long list of secondary plant compounds. Cyanide, nitrate, fluoroacetate, cyanogenic glycosides, saponins, oxalates, mimosine and various sterols are but a few. Quite recently, saponin concentrations have been implicated in low productivity in young cattle on young growth of signal grass (*Brachiaria*) (Lowe, S., personal communication). These may or may not be modified in the rumen by microbial action. The primary compound or its breakdown products in the rumen may be toxic or of no nutritional consequence. Some secondary plant compounds are actively detoxified in the liver.

One of the best known secondary plant compound in tree leaves is mimosine in *Leucaena*, which is degraded to the toxic compound 3-hydroxy-4-pyridone by normal rumen organisms and which is usually further degraded by another microbe where animals have evolved in grazing situations containing *Leucaena*. This microbe is not present in ruminants that have been isolated from *Leucaena* and inoculation is necessary to prevent toxicity when *Leucaena* becomes a high proportion of the total diet (Jones and Megarrity, 1986).

The main reason for the foregoing discussion is to emphasize that these secondary plant compounds should be taken into consideration. However, the toxic compounds often only become of significance nutritionally when the plant assumes a high proportion of the diet. Although some effects of the toxic compounds may persist when these are used to supplement the diets, the beneficial effects of a high protein forage often override their effects. There are, however, important exceptions particularly in the case of fluoroacetate which is an lethal compound that claims the lives of large numbers of cattle on pasture lands with *Acacia georginae*, even though it is of relatively low digestibility. Recently a rumen microbe has been genetically modified to carry and express a gene that encodes for an enzyme that hydrolyzes fluoroacetate (Gregg *et al.* 1994).

Comparisons of Tree Foliages and Multinutrient Blocks

Tree foliages have been given high prominence as protein supplements for ruminants fed low protein forages. However, seldom is it known whether escape protein *per se* is the valuable component of the foliage or whether the protein of the tree foliage is largely providing ammonia (from protein degradation), minerals, or all three. The rôle that tree foliages play in ruminant nutrition determines the required rate of supplementation, so it is extremely important to know what this rôle is.

Tree foliages low in tannins or other secondary plant compounds that might bind protein, are probably degraded rapidly and, at times, completely. In this way they provide ammonia and volatile fatty acids in the rumen. The rate of microbial growth on protein, however, is approximately half that on carbohydrate, so P/E ratios are lower when protein is degraded in the rumen in comparison with carbohydrates. On the other hand, if foliage protein is bound by condensed tannins, microbial degradation of leaf protein in the rumen will be prevented or slowed.

This will allow particles high in protein to move to the lower digestive tract where some of the condensed tannin complexed with protein may be hydrolyzed, which then allows the protein to be digested. Condensed tannin under acid or alkaline conditions of the intestines may be split to sugars and organic acids, mostly gallic acids, releasing protein and amino acids that are digestible in the lower gut. However, lowered N retention in animals fed some tanniniferous forages suggest that much bound protein, is unavailable for digestion in the gut (Mangan, 1988; Kumar and Singh, 1990; Waghorn and Shelton, 1995). Provided some protein remains soluble and can provide ammonia in the rumen and perhaps escape protein, some of the detrimental effects would be reduced. In this case the protein level relative to tannins would be the most critical factor.

Few research studies have defined the rôle required of tree foliages where their feeding objective is as supplements to low digestibility forage. The extent to which tree foliage protein is degraded in or escapes the rumen is extremely important. If the tree foliage protein is totally degraded then it provides only ammonia and minerals for microbial growth (but both these may be more easily and economically provided from other sources such as MUMB, chicken manure or litter from broiler production).

Tree foliages, on the other hand, often have higher digestibilities than the pasture forage available and thus provide a more energy dense feed, allowing a higher total feed intake. In a situation where the fermentable N and minerals can be provided more economically, say, as MUMB then the use of such tree foliages is counter-indicated. The potential value of the foliage may only be realized by harvesting and processing so that it contains a higher percentage of bypass protein.

4

Animal Nutrition and Feed Rations

Livestock production in all its ventures is a major source of incomes all over Kenya, from the most productive to nearly desert areas, And for all livestock owners livestock feeding and nutrition is a major concern. Inadequate nutrition is a major cause of low live-weight gains, infertility and low milk yields in dairy cattle. Also pig, chicken, dairygoat and many other livestock producers have expressed challenges in feeding their animals optimally. The following will explain the principles of animal nutrition and some examples of how to make home feed rations based on the types of feed available in major agroecological zone

THE FEED NUTRIENTS

Animal feed needs to contain various groups of nutrients, and the composition depends on which type of animal is being fed and the stage of production. Generally,

- Protein is the one element most necessary for body building and maintenance as well as milk production. Without protein there would be no body weight gain nor milk production
- Carbohydrates provide energy and body fat,
- Fats provide energy and the excess is converted to fat and stored in the body
- Minerals help in body building as well as in biological regulation of growth and reproduction. They are also a major source of nutrients in milk
- Vitamins help regulate the biological processes in the body and become a source of nutrients in milk
- Water helps all over in body building, heat regulation, biological processes as well as a large constituent of milk production as well as eggs.

When calculating feed needs of different animals a system called Metabolisable Energy as a basis for formulating rations on the farm is used. Metabolisable energy basically means that part of the feed which the animal is able to utilize.

The unit of energy in the ME system is the Joule (J) of which one million units (1000 000 J) is referred as Megajoule (MJ). All foods contain energy, but not all of it is available to the animal. Parts of all feed is lost in the faeces, this part is described as indigestible. Other losses of energy occur in the production of methane, the urine of the animal and the loss of body heat. The energy remaining after all this is called the Metabolisable Energy or ME.

Basically, feed organic nutrients are required by the animals for three things:

1. Use as materials for the construction of body tissues (growth and maintenance)
2. Synthesis of products such as milk and eggs
3. Use as sources of energy for work done. The work done include both metabolic (heat increment and maintenance) and physical *e.g.* walking and feeding.

USE OF FEED ENERGY WITHIN THE ANIMAL

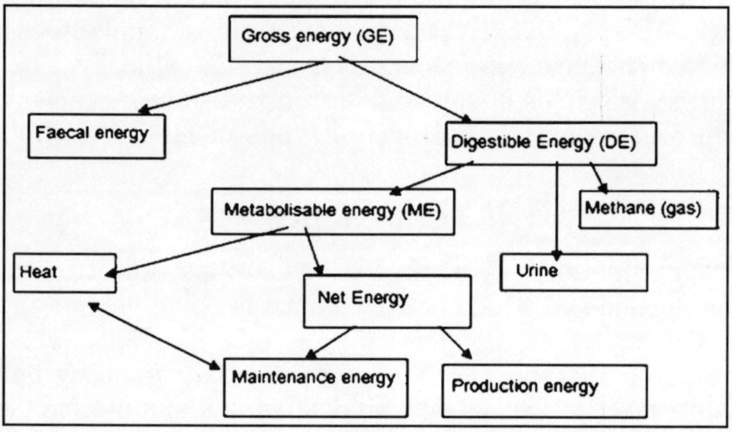

Gross Energy (GE)

The feed is comprised of chemical ingredients which are broadly classified as carbohydrates, proteins, lipids and vitamins. Heat is released when organic material such feed is burnt. For this reason, methods have been developed to measure the quantity of chemical energy present in a feed by determining the amount of heat generated from complete burning a known quantity. This is referred to as gross energy. Most of the common feeds have energy content of about 18.5 MJ/kg DM.

Digestible Energy (DE)

Not all the gross energy in consumed feed is available and useful to the animal. Some energy is lost from the animal though excretions and heat. The digestible energy is calculated by subtraction of feaces energy from gross energy. The DE represent the energy content of the digested nutrients

Metabolizable Energy (ME)

The animal further loses energy containing-substances through excretion of urine and production of gaṣes during metabolic processes. Metabolizable energy is what remains after subtraction of energy lost from urine and combustible gases resulting from the consumption of a feed. Loss of energy through methane (a combustible global warming gas) can be substantial, particularly for ruminants, hence can be of serious nutritive and environmental consequence.

Heat Increment (HI)

The ingestion of feed by an animal is also followed by losses of energy not only as the chemical energy excreta and gases produced but also as heat. Animals are continuously producing heat and losing it to their surroundings, either directly through radiation, conduction and convection or indirectly through water evaporation from the body. The heat is generated through processes of digestion and metabolism of nutrients derived from the feed. For instance, the act of eating, which includes chewing, swallowing and secretion of saliva, requires muscular activity and this generates heat. Unless the animal is in a particularly cold environment, this heat energy is of no value to it, and must be considered, like the energy of the excreta, as a tax on the energy of the feed. Energy lost in this manner is referred to as Heat increment.

Net Energy (NE)

The deduction of the HI of a food from its ME gives the Net energy, which is the energy available to the animal for useful purposes such as body maintenance and various forms of production.

Energy Feeds

Energy is the fuel that keeps all body functions working. Milk production requires a lot of energy. If energy in the ration is not enough, the animal will lose body condition and for milking cows, milk yield will drop, pregnant cows become ill after calving and the calf will usually be small in size. If there is excess energy in the ration, the animals becomes too fat. Cows that are too fat at calving usually have difficult births, often have problems with retained placenta, displaced abomasums and may suffer from milk fever and ketosis.

Sources of energy are roughages and concentrate supplements fed to your animals Minerals are required in small amounts than the other nutrients but are important components of the ration. They are essential for cows to remain healthy and for the body to function properly, for the development and maintenance of strong bones and for successful reproduction and production of milk and eggs. Roughages form the main bulk of the dairy cow ration. Roughages are bulky feeds that have a low weight per unit volume. Generally feedstuffs with more than 18 per cent crude fibre and low digestibility are considered as roughages. A high yielding cow may not have enough capacity to consume the amount of roughage required to supply sufficient energy required due to limitation of stomach size.For this reason, supplementing roughage diets with feeds high in readily available energy is often recommended. Examples of energy sources (Forages and fodders, agricultural by-products, and concentrates) are shown in Tables.

The currently recognized energy feed nutrients include:

- Carbohydrates such as Glucose, Fructose, Galactose, Sucrose, Maltose and Lactose, all different types of sugar
- Polysaccarides such as
  - Starch, found in roots and tubers as well as in grain,
  - Hemicellulose (somewhere between sugar and cellulose chemically speaking),
  - Cellulose, the principal constituent of cell walls of plants. Most abundant in more fibrous feeds, generally low in digestibility. Cattle, goats, sheep and horses digest cellulose fairly easily. Pigs and chicken do not digest cellulose very easily.
  - Lignin which essentially is not digestible to animals. Found in overmature hays, straws and hulls. High lignin content in feed may reduce the digestibility of cellulose and other nutrients.
- Fats and oils. Found in seeds, grains, avocadoes etc. Fats contain 2.25 times as much energy per kg compared to carbohydrates, but are usually expensive to produce.

Minerals

Minerals are chemical elements which form important component of animal feed ingredients. They are essential in ensuring normal and proper functions of the body as well as in maintenance of good health. When an element classified as essential lacks in the diet, the cattle will show deficiency symptoms, which are eradicated or prevented by inclusion of this particular element in the diet. Some elements are required in relatively large amounts compared to others. For this reasons the minerals have been classified as *'macro-minerals'* (required in larger amounts) and *'micro-minerals'* or *'trace-minerals'* (required in minute amounts).

Of the 20 elements that function in animal nutrition, carbon, hydrogen, oxygen and nitrogen are regarded as the non-mineral elements. The other 16 are referred to as the mineral elements which function in animal nutrition. Of these 7 are macro-minerals (required in fairly large amounts) and 9 are micro-minerals (required in very small or trace amounts). Micro-minerals are also sometimes called trace-minerals. Different livestock types have different mineral requirements, which as far as possible will be described under each livestock type. The macro-minerals are: Calcium, Phosphorus, Potassium, Sodium, Sulphur, Chlorine, Magnesium. The micro- or trace minerals are: Iron, Iodine, Copper, Cobalt, Fluorine, Manganese, Zinc, Molybdenum, Selenium.

Macro-minerals

Dairy cows require more of the macro-minerals (Calcium, Phosphorus, Magnesium, Sodium, Potassium, Chlorine, Sulphur) than the micro-minerals (Iodine, Iron, Cobalt, Copper, Manganese, Molybdenum, Zinc, Selenium).If cows do not consume enough of the macro-minerals, this will cause reduced milk yield, infertility problems, weakness of the bone and increased incidences of non-infectious diseases such as milk fever (Due to insufficient Calcium). Deficiencies in micro-minerals (Trace elements) can cause a variety of diseases and conditions depending on which mineral is deficient. Cattle grazing in areas around Nakuru usually have Cobalt deficiency and may develop a wasting disease called Nakuritis.

They become anemic and eventually die. The forages are deficient of mineral Cobalt because the soils naturally contain very low levels of this micro-nutrient. Special mineral supplements are available for cattle in such areas. Too much of the micro-minerals can cause poisoning. Calcium and Phosphorus are of particular importance when formulating rations. Legumes tend to have more Calcium and Phosphorus than grasses. Grains are low in Calcium. Young dark green forage tends to have more minerals than old, dry and yellow forages. Most tropical forages are low in Phosphorus. Extra Calcium and Phosphorus usually need to be provided in the ration over and above that naturally present in the feed and mineral mix, especially for high yielding animals. Tables 2 and 3 shows examples of sources of mineral salts (Forages and fodders, agricultural by-products, concentrates and minerals). *Salt*: (Sodium chloride) deficiency develops slow (weeks) but causes unthrifty appearance and low performance. Provision of ad lib salt licks are recommended. Plants tend to be low in both sodium and chlorine. It is therefore an important practice to give common salt to herbivores such as dairy cattle in order to prevent deficiency symptoms. Feeding diets deficient in salt may not show immediate symptoms, but chronic deficiency dairy cattle diets has been shown to lead to low appetite, low milk production and loss of weight. The addition of salt in the diet usually provide immediate cure.

Calcium: Calcium is the most abundant mineral in the animal body. It is the most important constituent of the skeleton (bones) and teeth. Calcium also plays important roles in the activities of enzymes and hormones, which catalyze and/or balance the body metabolic processes. Agricultural lime, fish meal, milk, crushed shells, marble dust, some seaweed and green leafy forages, especially legumes, are good sources of calcium. Calcium tend to be low in old, dry and yellowing forages.

In dairy cows, a condition known as 'milk fever' (*parturient paresis*) commonly occurs shortly after calving. It is characterized by a lowering of the blood calcium level (hypocalcaemia), muscular spasms, and in extreme cases paralysis and unconsciousness.

Deficiency symptoms: (a) rickets in young stock. Joints become enlarged. Bones become soft and deformed. Condition may be corrected in early stages with calcium feeding. (b) Osteomalacia or osteoporosis in older animals. Bones become porous and weak. Condition may be corrected by feeding calcium if bones do not break.Examples are known of cows fed too little calcium breaking their backs during mating

Phosphorus: Phosphorus is needed for bone and teeth formation, building body tissue (growth of animals), milk and egg production. Signs of phosphorous deficiency include animals eating soil, chewing on non-feed objects, slow or poor appetite, slow gain of bodyweight, low milk or egg production. Low dietary intakes of phosphorus have also been associated with poor fertility, apparent dysfunction of the ovaries causing inhibition, depression or irregular oestrus. Sources of Phosphorous: Bone meal, Rock phosphate, Superphosphates such as TSP etc. Also many improved salt licks contain phosphorus. Cereal grains are a good source of Phosphorous, but hays and straws have very low phosphorous content.

Magnesium: Magnesium is needed in proper functioning of the nervous system, carbohydrate metabolism and enzyme systems.

Deficiencies: (a) Hypermagnesemia also called grass tetany, grass staggers and wheat poisoning can occur when animals are grazing on young fresh grass or wheat heavily fertilized with nitrogen and with very little content of magnesium. Symptoms are hyper excitability and frequent death. More common in Europe than in Africa. Prevention: use animal salts containing magnesium especially when animals are grazing on new young grass or grains such as oats.

Sulfur: Sulfur requirements of cattle and sheep are around 0.1-0.2 per cent of ration dry matter. For non-ruminants sulphur should be in the form of sulfur-containing proteins. A deficiency of sulfur will express itself as a protein deficiency, general unthriftiness and poor performance.

Micro-minerals

Iron: Necessary for blood and some enzyme formation. The precise minimum requirements have not been determined for various classes of livestock, but 80mg of iron per kg of diet is more than adequate for most animals. Deficiencies are most often found in young pigs (other animals much less sensitive): Laboured breathing, Flappy wrinkled skin, edema of head and shoulders, pale eyelids, ears and nose. Prevention/cure: A few drops of ferrous sulphate or similar daily during the first 3-4 weeks. Salt licks containing iron.

Iodine: Needed for the production of Thyroxine in the thyroid gland. A level of 0.25 mg/kg air dried diet is considered adequate for most classes of livestock. Dairy cows should be provided with 0.5 mg iodine/kg dry matter feed. Deficiency symptoms: Goiter at birth or soon after, Hairlessness at birth, infected navels, dead or weak at birth. Prevention: mix normal iodized salt (table salt) into the salt licks of the livestock.

Cobalt: Needed in vitamin synthesis. For cattle and sheep, feed containing from 0.05-0.10 mg of cobalt/kg feed prevents any cobalt deficiency. For pigs cobalt is only needed as part of Vit B12. Several areas in Kenya have cobalt deficient soils, producing feed deficient in cobalt.particularly around Nakuru and Naivasha due to the absence of this element in the soils, leading to the absence in the pastures. A feed analysis will show whether feed in your area is cobalt deficient. Consult your livestock officer.

Deficiency symptoms are simply those of malnutrition: poor appetite, unthriftiness, weakness, anemia, decreased fertility, slow growth and decreased milk and wool production. There are number of disorders due to cobalt deficiency characterized by emaciation (wasting disease or *Nakuritis*), pining, anaemia and listlessness. Although excess cobalt can be toxic to animals, there is a wide margin of safety level. Thus cobalt toxicity is generally unlikely. Prevention and cure: Where cobalt deficiency is diagnosed, 12.5g of any cobalt salt, such as cobalt chloride, cobalt sulphate or cobalt carbonate can be mixed with 100 kg of normal cattle salt. Copper: needed for blood and hair production as well as in the enzyme system. Where diets are not high in Molybdenum and/or sulphate the following levels of copper per kilo of diet dry matter have been found adequate:

- *Dairy cattle*: 10 mg/kg
- *Beef cattle and sheep*: 4-5 mg/kg
- *Pigs*: 6 mg/kg
- *Horses*: 5-8 mg/kg

High levels of Molybdenum and/or sulphate may increase the copper requirements 2-3 times. Many areas in Kenya have copper deficiency in the soils and produce feed deficient in copper. Deficiency symptoms are not specific and may include any of the following: Bleaching of hair in cattle, abnormal wool growth in sheep, muscular incoordination, weakness at birth, anemia.Prevention and cure: Supplementation of livestock with copper in copper deficient areas is essential.

This can be done by using trace mineralized salt containing from 0.25-0.50 per cent copper sulphate. Pigs may be fed up to maximum 250g copper/kg dry feed. More than 100 mg copper per kilo dry matter may be toxic to cattle and over 50mg/kg will be toxic for sheep. It is also possible to repair your grazing areas for especially ruminants by upgrading the soil content of copper according to soil analysis recommendations. Generally grass and fodder deficient in copper have yellow or burnt leaf tips and low rates of production.

Fluorine: Necessary for healthy teeth, but excess may weaken and stain the teeth. In Kenya fluorine deficiencies are not common, but drinking water especially from boreholes often contain very high levels of fluorine. If the levels of fluorine are too high water can be filtered through a filter containing burnt bones, which will absorb most of the fluorine. This is more practical for human water consumption than for livestock. In Kenya the best advice for water treatment for excessive fluorine can be obtained from the Catholic Diocese in Nakuru.

Manganese: Influences estrus, ovulation, fetal development, udder development, milk production, growth and skeletal development. Requirements:

- *Dairy cattle*: 40 mg/kg of dry matter feed
- *Beef cattle and sheep*: 5-20 mg per kg dry matter feed
- *Pigs*: 10-20 mg/kg dry matter feed.

Deficiency symptoms noted from areas deficient in soil manganese include: delayed estrus, reduced ovulation, abortions, resorptions, deformed young, "knuckle over" in calves, poor growth. Supplementation is easily done with trace mineralized salts containing 0.25 per cent manganese.

Molybdenum: Important in poultry as it stimulates uric acid formation, and in ruminants stimulates action of rumen organisms. Molybdenum deficiencies have only been observed in poultry in special cases. Molybdenum supplementation is normally not recommended in livestock production.

Selenium: works in vitamin E absorption and utilization.

Requirements: about 0.1 mg or less per kg dry feed. Deficiency symptoms

Include: Nutritional muscular dystrophy in lambs and calves, retained placenta in cows, heart failure, paralysis, poor growth, low fertility, liver necrosis, pancreatic fibrosis in chicks. Many areas in Kenya are known to have selenium deficieny of the soils. If selenium deficiency is expected, a soil or feed sample can be sent to any of the major laboratories for analysis. Supplementation must be done very carefully as selenium in too large quantities is poisonous. 1 gram Selenium in the form of sodium selenite can be added to 10 kg dry feed in deficient areas (=10g or 2 teaspoons per 100kg feed-really not much).

Zinc: promotes growth and thriftiness. Promotes wound healing, related to hair and wool growth. Deficiencies mostly found in pigs fed on concrete floors. Deficiency symptoms include: general unthriftiness, poor growth, unhealthy looking hair, skin and wool, slow wound healing. Pigs can be supplemented with 50 mg of zinc per kg of dry feed or as trace mineralized salt.

FEED COMPOSITION FOR CATTLE AND SHEEP

Nutrition research spanning more than 100 years has defined the nutrients required by animals. Using this information, rations can be formulated from feeds and ingredients to meet these requirements. Animals fed these rations should not only remain healthy but be productive and efficient. The ultimate goal of feed analysis is to predict the productive response of animals when they are fed rations of a given composition. this is the real reason for information on feedstuff composition.

TABLE VALUES FOR FEEDSTUFF COMPOSITION

Feedstuffs vary in composition. Unlike chemicals that are "chemically pure" and therefore have a constant composition, feeds vary in their composition for many reasons. Actual analysis of a feed to be used in a ration is more accurate than tabular data. Obtain and use actual analysis whenever possible. Often, however, it is either impossible to determine actual compositional data, or there is insufficient time to obtain an analysis. Tabulated data are the next best source of information. When using tabulated data, remember that feeds vary in their composition. the organic constituents (*e.g.*, crude protein, ether extract, crude fibre, acid detergent fibre and neutral detergent fibre) can vary as much as 15 per cent, the mineral constituents as much as 30 per cent, and the energy values at least 10 per cent, from table values.

Therefore, the values shown can only be guides. For this reason they are called "typical values." they are not averages of published information. Some judgement was used in arriving at some of the values in the hope that the values will be realistic for use in cattle and sheep rations. Feeds can be chemically analyzed for many things that may or may not be related to the response of the animals to which they are fed. In the accompanying table, certain chemical constituents are shown. the response of cattle and sheep when fed a feed, however, can be termed the biological response to the feed in question. this is a function of its chemical composition and the ability of the animal to derive useful nutrients from the feed. The latter relates to the digestibility or availability of a nutrient in the feed for absorption into the body and its ultimate efficiency of use in the animal. this also depends on the nutrient status of the animal and the productive or physiological function being performed by the animal. Ground fence posts and shelled corn may have the same gross energy value in a bomb calorimeter, but have markedly different useful energy value (TDN, digestible energy, net energy) when consumed by the animal. That means that the biological attributes of a feed have much greater meaning in predicting the productive response of animals. However, they are more difficult to accurately determine because there is an interaction between the chemical composition of the feed and the digestive and metabolic capabilities of the animal being fed.

USING INFORMATION

Feed Names

The most obvious or commonly used feed names are given in the table. Feeds designated as "fresh" are feeds that are grazed or fed as fresh cut materials.

Dry Matter

Typical dry matter (DM) values are shown. However, the moisture content of feeds can vary greatly. thus DM content can be the biggest reason for variation in the composition of feedstuffs on an "as fed basis." For this reason, the composition of chemical constituents and biological attributes of feeds are shown on a DM basis. Because DM can vary greatly, and because one of the factors regulating total feed intake is the DM content of feeds, ration formulation on a DM basis is more sound than using "as fed basis." To convert the values shown to an "as fed basis," multiply the decimal equivalent of the DM content times the compositional value shown in the table.

Protein

Crude protein (CP) values are shown for each feed. Crude protein is determined by taking the Kjeldahl nitrogen times (100/16 or 6.25). (Proteins contain 16 per cent nitrogen on average.) Crude protein does not give any information on the actual protein and non-protein content of a feed. Digestible protein has been included in many feed composition tables, but because of the large contribution of body protein to the apparent protein in the feces, digestible protein is more misleading than CP.

Calculate digestible protein from the CP content of the ration fed to cattle or sheep by the following equation: per cent DP = 0.9 (per cent CP) - 3, where per cent DP and per cent CP are the ration values on a dry matter basis. Rumen "by-pass" protein, or undegraded intake protein (UIP), represents the per cent of protein that passes through the rumen without being degraded by rumen microorganisms. Like other biological attributes, these values are not constant. By-pass values for many feeds have not been determined. Reasonable estimates are difficult to make.

Degradable intake protein (DIP) is used to meet the nitrogen requirements of rumen micro-organisms. Nitrogen sources such as urea are the most economical sources of DIP. Balancing DIP and UIP sources provides a more accurate way of meeting the metabolizable protein needs of ruminants.

Crude, Acid Detergent and Neutral Detergent Fibres

After more than 100 years, crude fibre (CF) is declining in popularity as a measure of low digestible material in feeds. the major problem with CF is that variable amounts of lignin, which is not digestible, are removed from various feeds in the CF procedure. In the old scheme, the material removed was called nitrogen-free extract (NFE) and was thought to be more digestible than CF, even though many feeds have been shown to have a higher digestibility for CF than NFE. One reason CF remained in the analytical scheme for feedstuff analysis was its requirement for the determination of TDN. Newer procedures have developed an alternate analytical scheme, namely, acid detergent fibre

(ADF) and neutral detergent fibre (NDF). ADF is highly related to digestibility in the animal. NDF is related to voluntary intake of the feed and the availability of net energy from digestible energy. Both measures relate more directly to predicted animal performance, so they are more valuable than CF. Also, if TDN is replaced by other measures of energy value, there will be little use of the CF content of feeds. As more complete data on the ADF and NDF content of feeds are developed, CF will be dropped.

Minerals

Values are shown for only certain minerals. Calcium (Ca) and phosphorus (P) are important minerals in most feeding situations. Potassium (K) becomes important as the level of concentrate increases in the ration, or when non-protein nitrogen is substituted for intact protein. Sulfur (S) also becomes more important as the level of non-protein nitrogen increases in the ration.

Vitamins

Vitamins have been omitted from the table. Only vitamin A is of general practical importance in cattle and sheep feeding. the vitamin A and carotene in feeds depend largely on maturity and conditions at harvest and the length and conditions of storage. therefore, it is probably unwise to rely entirely on harvested feeds as a source of vitamin A. Where roughages are being fed that contain good green colour or are being fed as immature fresh forages (*e.g.*, pasture), there will probably be sufficient vitamin A.

Energy

Four measures of the energy value of feeds are shown in the table. TDN is shown simply because there are more TDN values for feeds, and because this has become a standard system for expressing the energy value of feeds for cattle and sheep. there are several technical problems with TDN, however. there is a poor relationship between crude fibre and NFE digestibility in certain feeds. TDN also overestimates the value of roughages compared to concentrates in producing animals. Some have argued that energy is not measured in pounds or per cent, so TDN is not a valid measure of energy. However, this is more a scientific argument than a criticism of the predictive value of TDN.

Digestible energy (DE) values also are shown. Many studies have shown there is a constant relationship between TDN and DE: there are 2 Mcals of DE per pound of TDN. Obviously, DE can be calculated by multiplying.02 times the per cent TDN content. Because DE is measured in calories, it is technically preferred over TDN. With greater emphasis on ADF and NDF as replacements for CF and the use of the bomb calorimeter to measure DE directly, use of TDN should gradually decrease. It should be apparent, however, that the ability of TDN and DE to predict animal performance is exactly equal.

Interest in the use of net energy (NE) in evaluating feeds for cattle and sheep was renewed with the development of the California net energy system. the main reason is the improved predictability of results depending on whether feed energy is being used for maintenance (NEm) or growth (NEg). the major problem in using these NE values is predicting feed intake and, therefore, the proportion of feed that will be used for maintenance and growth.

Some use only the NEg values in formulating rations. this suffers the equal but opposite criticism mentioned for TDN — NEg overestimates the feeding value of concentrates relative to roughages. Others use the average of the two NE values, but this would be true only for cattle or sheep eating twice their maintenance requirement.

FEED FORMULATION

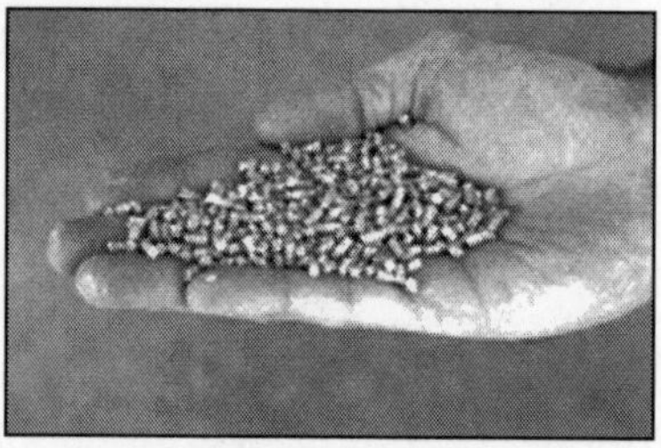

Fig. Pelleted Poultry Feed

Feed formulation is the process of quantifying the amounts of feed ingredients that need to be put together, to form a single uniform mixture (diet) for poultry that supplies all of their nutrient requirements. It is one of the central operations of the poultry industry, in view of its role in ensuring good nutrition. Feed costs account for more than 70 per cent of the total production costs for most types of poultry, so it is important that returns are maximised through use of adequate diets. Feed formulation is a central operation in poultry production, ensuring that feed ingredients are economically used for optimum growth of chickens.

It requires a good knowledge of poultry and feed ingredients. Most large-scale poultry farmers depend on commercial feed mills for their feeds, to obviate the need to do their own formulations or feed preparation. It is therefore essential that formulations are accurate, to ensure a large number of flocks are not adversely affected.

Typical Formulation

Feed formulation is both a science and an art, requiring knowledge of feed and poultry, and some patience and innovation when using formulae. Typical formulations indicate the amounts of each ingredient that should be included in the diet, and then provide the concentration of nutrients (composition) in the diet. The nutrient composition of the diet will indicate the adequacy of the diet for the particular class of poultry for which it is prepared (*e.g.* egg layers, meat chickens or breeders).

It is common to show the energy and protein contents of the diet but comprehensive information on concentrations of mineral elements and amino acids may also be provided. Since lysine is a key amino acid for poultry, the concentrations of the other amino acids may be related to it. Because some nutrients interfere with the utilisation of other nutrients, relationships between nutrients, particularly amino acids, also show if each of the nutrients will be used efficiently.

One of the reasons for formulating diets for poultry has already been identified, *i.e.* to produce a single uniform feed that can be delivered efficiently to poultry. The production rates expected on modern poultry farms also dictate that the dietary requirements of poultry be carefully identified and precisely met. Diet formulation enables the poultry industry to maintain some uniformity in levels of production. Industry nutritionists tend to use the same formulations over relatively long periods of time, so that the quality of the product (meat or eggs) remains stable over time. The product quality can also be easily predicted if the same diet formula is used and all other factors remain unchanged.

It is worth stating that chickens are able to select from different ingredients, placed separately, in order to grow and lay normally. Such a practice is known as choice feeding, and is practised on a limited scale by some poultry producers. However, choice feeding is cumbersome and cannot be economically applied in large-scale modern commercial situations without modifications to the infrastructure.

FEEDING STANDARDS AND FEED EVALUATION

To ration concentrates effectively to the individual cow, the dairyman needs to know (a) how much TDN the cow needs and how much can be fed profitably, (b) the TDN value of the forage to be fed, and (c) how much of the particular forage will be ingested. Thus, effective rationing requires a satisfactory feeding standard and information on the over-all nutritive value of the particular forage (s) to be fed. A limitation in either the feeding guide employed or the appraisal of the nutritive value of the forage naturally is reflected accordingly in the amount of concentrates fed.

Recent attempts have been made to develop rapid and inexpensive, but indirect, methods for the assessment of the nutritive value of forage in programmes which have become known as forage testing. The ultimate aim in the use of a forage-testing programme is that of determining the kind and amount of concentrates that should be included in the ration of the individual cow as dictated by the feeding standard. Although forage-testing schemes in their present form and manner of application provide an estimate of the TDN value per unit weight of forage, they do not provide an estimate of the more influential determinant of the total energy intake, the expected forage intake. As a consequence of this limitation, and others which have been detailed in the text, it is concluded that forage-testing schemes contribute very little information of genuine value to the daily rationing and the improvement in milk yield of the individual cow. This is particularly true when the shortcomings of forage testing are considered in relation to the limitations of existing feeding standards and certain commonly encountered defects in practical feeding.

Most forages consumed by livestock in developing countries have a low digestibility which rarely exceeds 55 per cent and is mostly in the range of 40–45 per cent. The calculated metabolisable energy in the dry matter (ME/DM), thus, ranges from 7.5 MJ down to 4.8 MJ. Feeding standards indicate that feeds with a metabolisable energy content, of 7.5 MJ will support growth rates of cattle of approximately 2 g/MJ of ME intake. On a forage at the lowest level of ME, cattle would be in negative energy balance.

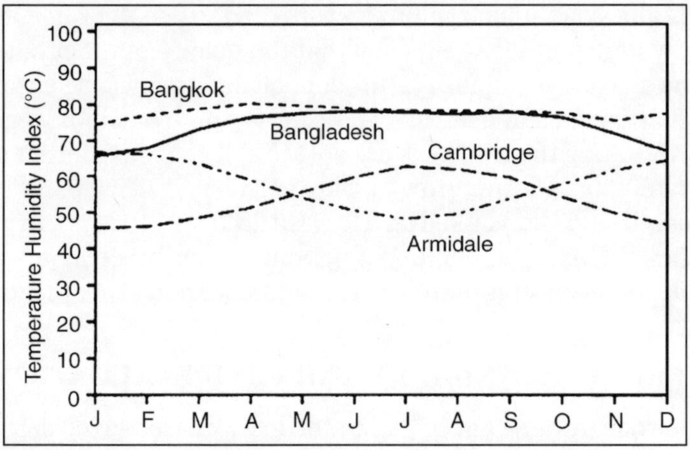

Fig. Temperature Humidity Index (THI) of Climates in Temperate Countries as Indicated by Cambridge (U.K.) and Armidale (Australia) as Compared to Tropical Countries as Indicated by Bangkok (Thailand) and Mymensingh (Bangladesh). The THI is Calculated on the Mean of the Maximum-minimum Temperatures and Humidities THI(°C) = Temp. (dry bulb) + 0.36 Temp. (dew point) + 41.2°C

Contrast this with results of supplementary feeding trials based on balancing the nutrition of animals with urea/minerals and bypass protein, where cattle growth rates equivalent to 18 g/MJ of ME intake have been achieved in cattle fed straw. Obviously the presently accepted feeding standards have been very misleading and can not be used as a means of predicting animal performance. Of vital importance however, is that the application of the concept of balanced nutrition can improve animal growth by 2–3 fold and the efficiency of animal growth by as much as six fold over previous estimates (a range of 2–10 fold). In addition it also shows that although growth rates of cattle are below those on grain based diets cattle on forage based diets can highly efficiently convert feed to product.

Implications of Low Productivity of Ruminants in the Tropics

Low productivity of ruminant livestock has been accepted in developing countries as an inevitable result of the poor feed base and a low feed conversion efficiency. The concept being that there is a large heat production (energy requirement) associated with the ingestion, movement of digesta along the tract in animals on forages as compared to concentrates. This conclusion is contrary to the conclusions of Leng and the concept of balanced nutrition presented here.

SOME BASIC EXPLANATIONS FOR THE INEFFICIENCY OF RUMINANTS ON FORAGE DIETS

Inefficiency of Acetate Utilisation

The original calorimetric studies of Graham and his colleagues indicated that infused acetate or butyrate were utilised by sheep with low efficiencies, *i.e.* there was a high heat increment when acetate was given compared with propionate. Considerable effort has since been expended on testing the hypothesis (or disproving it) that acetogenic substrate is used wastefully.

Blaxter and his colleagues used diets based on dried grass which was chopped and cubed. It had a metabolisable energy content (M/D) of about 8.5 MJ/kg dry matter. As most of the protein in the diet could have been highly soluble, the P/E ratio in absorbed nutrients would have been relatively low. The controversy concerning the efficiency with which acetogenic substrate is utilised may be rationalised at least to some extent by considering the balances of nutrients available to the ruminants in the various experiments and the ability or otherwise of the animals to synthesise fat, dissipate heat or to oxidise substrate to keep warm.

For example the presence of small amounts of fish meal, that has a considerable amount of protein that escapes the rumen, in a concentrate diet provides an explanation for the differences between these authors' results and those summarised by Blaxter where sheep were fed dried grass which may have contained a highly soluble source of protein.

Requirements for Glucose by Ruminants

The need to manipulate or supplement diets for ruminants in order to ensure an adequate supply of glucose and of glucogenic compounds was discussed fully by Preston and Leng, who made a comprehensive literature survey. In outline, the rationale that is used to justify the concept of glucose being a limiting nutrient is as follows: Little glucose is absorbed by ruminants but they synthesise considerable glucose from precursors such as propionate and certain amino acids, largely in the liver. Glucose is certainly required, by ruminants, as a major substrate for cell synthesis, as an important oxidative energy supply in the brain and red cells and as an important nutrient for the growing foetus and for milk lactose and fat synthesis in lactating animals. Glucose needs to be oxidised in the adipose tissue, and to a lesser extent the mammary gland, to supply the reduced cofactors for fat synthesis from acetate (in this case of NADPH, generated in the pentose phosphate pathway).

It is possible therefore that fat synthesis in an animal may be limited by glucose availability for oxidation. If there is a block in acetic acid utilisation, it cannot be allowed to accumulate in the blood as this would lead to acidosis, the animal therefore, needs to oxidise the excess acetate in a futile cycle in which acetyl CoA is produced with a requirement for ATP and then hydrolysed. Alternatively an animal could increase its muscular activity by standing instead of sitting or walking long distances which would ensure increased acetate oxidation.

The animal that is most at risk to a deficiency of glucose is the animal with a big demand for glucose (late pregnant or early lactating) fed forage based diets (high acetogenic fermentation in the rumen) with little bypass nutrients in the diet (receives only amino acids from microbial sources) and is tethered (no requirements to oxidise acetogenic substrate to walk) living in a tropical country (no requirement to oxidise acetate to keep warm).

As discussed above, reduction in the need for acetate oxidation and a high glucose demand can establish a situation where a deficiency of glucose could lead to an imbalanced nutrition with a need to burn off excess acetate with a high thermogenic effect. Where such a thermogenic effect cannot be tolerated (high environmental temperatures, high humidity and perhaps heavy insulative coat) the animal needs to respond by reducing its feed intake. Conversely the thermogenic effect may be prevented by balancing the diet with nutrients that supply glucogenic substrates for absorption from the intestine.

Balancing Nutrition for Reproduction /Pregnancy and Lactation

Considerable research has indicated that the level of protein nutrition considerably influences reproduction of both male and female animals and subsequent pregnancy and lactation. Leng *et al.* indicated that on forage based diets each physiological function is affected adversely by a low P/E ratio in the nutrients absorbed by ruminants; this could be overcome, to some extent, by feeding protein meals which bypass the rumen.

In summary, feeding a supplement which improves the P/E ratio in the nutrients absorbed by ruminants on a low true protein, forage diet has the following potential effects, particularly in a hot climate, on reproductive efficiency of female ruminants:

- Stimulates liveweight gain of dam or reduces liveweight loss
- Improves ovulation rates

- Improves placental size
- Improves birth weight and so increases survival and because of the increased birth weight possibly lowers the incidence of retained placenta
- Increases milk yield and efficiencies of milk production.

Prevention of protein deficiency in early life also prevents stunting of final body size. Differences in size of animals of the same breed, in the same country, is almost always a result of differences in nutrition and not inherent differences. This has recently been emphasised with N'Dama breed which has always been considered to be a small breed weighing up to 250 kg liveweight. With good nutrition and adequate management the bulls have now been shown to grow to 500 kg liveweight.

Work from Nigeria and Australia has also shown that young and old bulls are also very susceptible to low P/E ratios in the nutrients absorbed. Young bulls that grow on diets that would have given a high P/E ratio in the nutrients absorbed, compared with animals fed a diet giving a low P/E ratio, had better testicular development and produced larger ejaculates with double the sperm content. Older bulls that go through a period of protein undernutrition have decreased testicular size and probably are less fertile.

IMPLICATION OF PARASITE/DISEASE AND NUTRITION

Undoubtedly any parasitic or disease condition that drains protein from the animal will increase the animals requirements for protein relative to energy. Similarly, infective agents that utilise glucose may also increase the demand for this critical nutrient.

For example, trypanosomes and epyrythrozoan parasites which invade red cells, increase protein requirements by increasing red cell turnover rate and also increase the animals requirements for glucose as this is the major substrate used in the parasite's metabolism.

It is suggested that improving the protein nutrition of ruminants through providing bypass protein directly to the animal (*i.e.* avoiding rumen fermentation) may considerably ameliorate the detrimental effects of intestinal and blood parasites and may assist in development of early immunity. (J. Steele personal communication)

IMPLICATIONS OF AN INCREASED NUTRIENT REQUIREMENTS FOR WORK

Light work requires acetogenic substrate probably acetate for muscle contraction but heavy work is probably dependent on long chain fatty acids mobilised from adipose tissue. All working muscles use some glucose and there is always a concomitant increase in glucose utilisation when free fatty acid metabolism is increased by the work load.

Imbalanced diets fed to draught animals which result in a low P/E ratio in the nutrients absorbed are not so disadvantageous as they are for a lactating or growing animal. The excess acetogenic substrate is used for work removing the necessity to otherwise enzymatically dispose of the substrate. However, the increment in requirements for glucose oxidation in muscles although small could be a limitation. In addition the need in heavy work to supply long chain fatty acids which can be oxidised to produce ATP at a faster rate than from acetate, indicates that there is an increased need for glucose not only in muscle metabolism but also to aid fat synthesis from acetate in the period of rest.

An imbalanced feed (low P/E ratio or low G/E ratio), which is associated with a high metabolic heat production, could also reduce the recovery of body temperature of draught animals when the animal is resting and, thus, reduce feed intake. A decreased fat deposition because of such an imbalanced diet during the non- working season may also be a major constraint to draught capacity because of lack of body reserves for mobilisation in the working season.

The P/E ratio required to support fat deposition (in periods of low work load) or to reduce heat stress by reducing metabolic heat is likely to be much lower than for example for growth and milk production. A draught animal fed poor quality forage probably requires little more nutrients than can be supplied from an efficiently functioning rumen. This is attained for example by supplying the animal with a molasses/urea multi-nutrient block.

Implications for Areas of Research

This discussion highlights the manipulable aspects of ruminant nutrition, where substantial improvements in productive efficiency of animals fed largely on fibrous feeds could result.

The differences between temperate and tropical conditions also indicate opportunities which are often not apparent in temperate countries and mitigates against direct transfer of results from the former to the latter. In tropical conditions, protein nutrition of ruminants is more crucial than in temperate areas.

The general conclusions are that:

- The primary constraint to ruminant production from fibrous feeds (in the tropics) is the low efficiency of feed utilisation often coupled with an environmental/physiological effect which results in a reduction in potential feed intake.
- In practice the most effective mechanisms for improving productivity will be to improve the p/e ratio in the nutrients absorbed because of the large effect on efficiency of feed utilisation.
- Draught animals will also benefit from technologies that increase microbial growth efficiency in the rumen and at times improve digestibility, intake and p/e ratio in the nutrient absorbed.

- Contrary to many statements in authoritative reviews, only when p/e ratios have been optimised does the energy density of the feed become a primary constraint.
- As the primary constraint is an inefficiency due to a low p/e ratio in the nutrients absorbed it is unlikely that other inputs [*e.g.* agents to repartition nutrients or to stimulate growth (B-agonists, BSt)] could be successful without first applying strategic supplementation.

FEED FORMULATION FOR OPTIMUM PROFITABILITY

Nutrition plays a major role in the profitability of animal production. Feed accounts for up to 70 per cent of total poultry and swine production costs. The goal of a feeding programme is to meet the nutrient needs of animals, based on optimal performance, with a mixture of ingredients, at the lowest cost. Achieving this goal requires a very complex process.

The purpose of this article is to discuss the important elements of establishing the nutrient needs of animals and how to match those nutrient needs with available ingredients. The ultimate goal is to optimize profitability by matching established nutrient needs of the animal with appropriate ingredient mixtures. The nutrient needs of an animal can be depicted by the following graph.

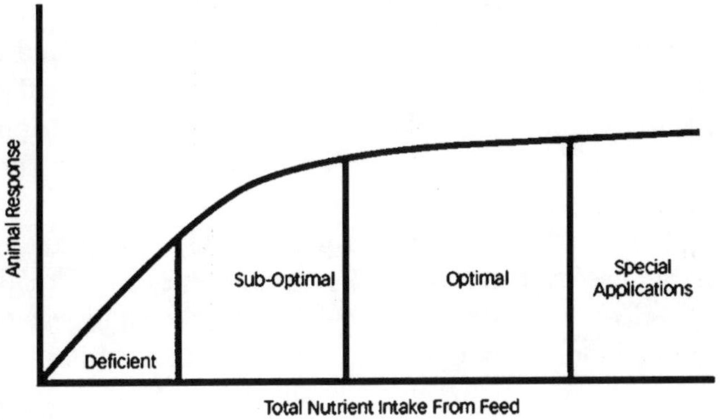

The goal of a feeding programme is to meet the nutrient needs of animals, based on optimal performance, with a mixture of ingredients, at the lowest cost. At low levels of nutrient intake, animals will exhibit specific deficiency symptoms based on the specific nutrient(s) that are deficient.

In this stage, animals are not receiving a diet adequate to support normal growth or productivity. No animal should ever be fed a deficient diet. A feeding programme in the sub-optimal stage will prevent deficiency symptoms, but the animal still is not receiving adequate nutrition to maximize growth, egg production, and/or profitability. The National Research Council publishes nutrient requirements for different species. These estimates are determined by researchers who critically evaluate the scientific literature and determine "best" estimates of nutrient requirements. These estimates are minimum standards and are not to be considered recommended nutrient allowances.

This is mainly because the estimates include no margin of safety margin. The stage of optimum nutrition typically optimizes profitable production. In this stage, animals are receiving nutrient levels sufficient to maximize productivity profitably. Nutrient requirements in this stage are best determined through sound production research. The law of diminishing returns would suggest animals should be fed just before maximized performance in this "optimum" stage to achieve maximum profitability. Often times this stage of nutrient levels has been historically used to provide a safety margin to account for variability in animals, feed ingredients, etc. However, as ingredient costs increase, the acceptable safety margin continues to be challenged.

Beyond this optimal stage of nutrition is a stage for special applications. Animal under immune challenges or applications where animals are being fed to meet a specific market demand may require nutrient levels in this stage. Otherwise, animals fed nutrient levels in this stage are only improving performance minimally or not at all. If there is not a specific need for nutrients at this level, feed cost is increased with no beneficial financial return.

Feed is made up of several ingredients, which are grouped into ingredients providing energy (fats, oils and carbohydrates), protein (amino acids), vitamins, and minerals. Cereal grains such as corn, wheat, and barley, plus additional fat will be used to mainly provide energy. Soybean meal, extruded and expelled soybeans, canola meal and poultry byproduct meal are used in diets to mainly provide protein. Ingredients must be evaluated for their nutrient content to help establish the amount of each ingredient to be included in the diet.

Feed ingredients are chemically broken into the different components described above. However, to determine the suitability of ingredients for poultry and swine, it is necessary to break these components into smaller fractions. For example, proteins are made of amino acids. Poultry and pigs don't actually have a requirement for protein, but there are ten amino acids required to be in the diet at specified minimum amounts. In this case, it is important to analyze the amino acid composition of the feed ingredients.

Likewise, ash is composed of several minerals (calcium, phosphorus, sodium, chloride, etc.), which are critical nutrients required at specific amounts in poultry and swine diets. Analysis of ash alone gives very little information about the suitability of an ingredient to meet the nutritional requirements of the animal. More detailed analysis of the actual minerals making up the ash component is needed. In addition, the extent of how well these nutrients are utilized by the animal (usually reported as digestible or available nutrients) can also be utilized in feed formulation and are determined through extensive research.

Chemical methods used to analyze ingredients are expensive and cumbersome, often requiring days to weeks to complete. However, newer equipment and procedures have been developed, which enable the rapid evaluation of most materials. For example, near-infrared reflectance spectroscopy (NIR) is one technique used by Wenger Feeds to quickly evaluate feed ingredients for amino acids, protein, fibre, and fat.

Nutrient content of ingredients can vary significantly from one supplier to the other. For this reason, it is important to determine the nutrient content of an ingredient by supplier. Nutrient content of an ingredient can also vary by season. Cereal grains like corn fill out ears poorly in drought years, which can reduce corn quality. New crop corn and soybean meal derived from new crop soybeans typically have a different nutrient content than corn and soybean meal from a previous year's crop.

Feed ingredients used in animal feeds must meet all guarantees and any pre-determined buying specifications. These guarantees include both physical and chemical parameters. The physical evaluation provides preliminary information on the quality of the ingredient. This involves assessing physical qualities such as weight, colour, smell, and whether the ingredient is contaminated with foreign material. This physical evaluation typically occurs in the ingredient receiving area of the feed mill and is a key step in the quality evaluation of a feed ingredient.

Although the average composition of many common ingredients is known, ingredients can be variable. For this reason, the average composition of ingredients is typically discounted to account for the degree of ingredient variability. This helps to meet targeted nutrient needs, regardless of ingredient variability. Nonetheless, finished feeds should be analyzed on a routine basis to verify that the process used for ingredient analysis results in the expected final feed.

The most important area to evaluate the suitability of an ingredient testing programme is in animal productivity. When feed is given to animals, they are only able to break down part of the feed and absorb it into the body for maintenance, growth, egg production, lactation, etc. The rest is lost through feces and urine, which are excreted together by poultry. Animal productivity defines the real nutritive value of the feed and must be measured and utilized as part of an ingredient and animal nutrient needs evaluation.

Ingredients must also be researched to determine optimum levels in feed. Many byproducts can be potentially used to reduce feed cost, but we must understand the influence of using the byproducts on animal performance. Often byproducts with proper nutrient values can be formulated into feed and result in no change in animal performance. Typically, there is a maximum inclusion level at which higher inclusion levels result in reduced animal performance and/or the ingredient is no longer cost effective to use. These maximum inclusion levels can also vary by productive stage (*i.e.* maximum inclusions might be lower in starter diets than finishing diets).

Wenger Feeds has research capabilities to evaluate nutrient requirements and ingredients for layers, pullets, broilers, and grow-finish swine in production settings. These research locations are designed to test nutrient levels and ingredients in a "controlled" research setting but still in facilities similar to those found in actual animal production. Examples of research conducted in these facilities include: an evaluation of maximum inclusion levels of Distiller's Dried Grains with Solubles (DDGS) in layer diets, meeting the body weight goals in pullets, estimating lysine and energy requirements of grow-finish pigs, and the influence of nutrition on skin and feather quality in broilers.

In summary, nutrient levels less than targeted levels can result in suboptimal animal performance, and nutrient levels in excess of targeted levels add unnecessary costs. Both result in reduced profitability. The goal in formulation is to meet targeted dietary nutrient needs with sound, cost effective ingredients critical to meet financial goals.

SOURCES OF DATA ON FEEDSTUFFS COMPOSITION

Information on feedstuffs composition and nutritional value plays a fundamental role at the farm, feed manufacturer and government levels, so the data inserted into databases need to be reliable and of high quality. The quality of the data depends, in the first place, on the origin of the information, which may have been extracted from the literature, collected from laboratories, obtained from other databases or generated specifically from animal studies and analyses of feeds.

DATA ASSEMBLED FROM THE LITERATURE

Feedstuffs information in the literature usually concentrates more on the data value obtained and does not give a full description of the feedstuffs concerned. When feedstuffs are insufficiently or inappropriately described there is a risk of confusing their real origin. A typical example of this occurs when publications generally refer to the product "distillers' dark grains", which is a by-product of the whisky distilling industry that, depending on the type of distilling process, may be derived from barley (in the malt distilling process) or wheat and/or maize (in the grain distilling process).

In this case a true and existing genetic variability is ignored. It is evident that the compositions of the two products will differ, so the provision of additional information on the origin of the feedstuff is essential; it is not enough merely to refer to it as "distillers' dark grains". In the absence of additional information, when trying to obtain information on the protein content specifically of barley distillers' dark grains, a value obtained from the average protein content of samples of both barley dark grains and wheat/maize dark grains would be retrieved from the database, and this value would be meaningless.

Data obtained from the literature can therefore often be inappropriate, particularly when they use averaged data which may incorrectly merge different feedstuffs. Data from the literature are recommended only when no other sources are available.

Data collected from laboratories

When compiling databases of feedstuffs information, the data used have usually been produced in a number of different laboratories. When assembling collections of data, it is necessary to have specific information regarding the analytical procedures used to obtain the data, in order to guarantee that those used to create a database have been obtained through standardized methods, are expressed in standardized units and can, therefore, be compared. The standardization issue is particularly important for data on the nutritional value of feedstuffs, obtained through *in vivo* digestibility or *in situ* degradability studies. In this case, together with the strictly analytical issue, the laboratories and research institutes involved should have standardized procedures for conducting animal studies. These relate to the numbers of animals and the species used (sheep, cattle, pigs, etc.) and to the feeding and collection protocols employed. However, when conducting *in vivo/in situ* studies, results will always be substantially influenced by variability among individual animals, no matter what level of standardization is achieved.

The standardized analytical procedures used to collect data from laboratories need to be monitored continuously, and the data they produce checked for accuracy and relevance. For this purpose, ring tests among participating laboratories should be carried out in order to evaluate the precision of standard or newly developed techniques.

Merging of New and Old Data

It is common practice to merge new databases with old ones. In these cases it is important that the analytical procedures used to obtain all of the data are understood in order to generate information that is expressed on a common basis and is, therefore, comparable. The evolution of analytical methodology is a substantial issue. Old data may have been obtained with obsolete methods or with surviving analytical procedures that, in the past, produced data that were less accurate than those produced today. Thus, the merging of old data with new may be inappropriate and may reduce the overall quality of the information.

Data Produced Specifically for Use in Databases

The data generated specifically for use in databases are likely to be those of the highest quality as all the necessary information will be included and the methods standardized and checked. The chemical composition of feeds can be analysed using the most up-to-date and standard methods and animal studies can be designed appropriately, by deciding which animal species to use and in which physiological state the animals studied should be. Both feeding and collection protocols can also be set with the aim of meeting the specific requirements of the database.

Ideally, an entire database would be produced by analysing specifically the composition of feedstuffs and studying their nutritional value through animal studies. In this way the variability arising from extrinsic factors such as analysis methods and animal protocols would be reduced to the minimum. On the other hand, when using data from a single source it is critical that the techniques and procedures used have been certified for precision and are representative. However, if it is to contain representative information, the database has to be established with an appropriate number of samples. The costs involved in analysing feedstuffs and, particularly, performing animal measurements are so prohibitive that it is rare for feedstuff information to be produced specifically for the purpose of assembling a database.

Moreover, much time is required to obtain the data, particularly those derived from animal studies. Therefore, the data are usually extracted from the different sources listed previously. Ultimately, it is the responsibility of the database manager to judge which information to include in order to generate a meaningful database.

SOURCES OF DATA VARIABILITY ON FEEDSTUFFS

Information on feedstuffs may be associated with different sources and types of variability. Variability can be "intrinsic" and caused by real differences among feedstuffs, such as the genetic origin of the feed or the processes that have been applied to it. Variability can also be "extrinsic", that is caused by differences in sampling and analysis procedures.

Intrinsic Variability of Data

Each feed has particular chemical and nutritional characteristics which distinguish it from other feeds. However, the same feed can also be derived from different cultivars or influenced by its geographical origin. Moreover, an identical raw material may have been processed in a variety of ways to produce very different products. The wide range of diverse wheat-related materials provides a clear illustration of the need to distinguish among different feedstuffs derived from the same raw material. Terms for wheat-based materials include wheat, durum wheat, winter wheat, spring wheat, processed wheat meal and wheat distillers' dark grains. All of these products and by-products are wheat-based, but they vary substantially in terms of both chemical composition and nutritional attributes and it is therefore expected that they will have different effects when fed to animals.

When assembling a database it is necessary to make sure that the real variability arising from the intrinsic characteristics of the feedstuffs is correctly recognized in order not to affect the quality of the data. Information from feedstuffs databases is most commonly retrieved in the form of averages (or "consolidated data"), so the feeds have to be merged judiciously in order to obtain meaningful information. To achieve reliability, a precise description and the correct naming of feedstuffs are essential, and for this purpose all available information on the type, origin, processing, etc. of the feedstuffs is needed.

One important attribute that is often omitted is information on the forage's stage of maturity. As shown in the example illustrated in the Figure, the content of both neutral detergent fibre (NDF) and starch relate strongly to the stage of maturity of the forage maize, and this is clearly reflected in the dry matter (DM) content of the plant, which increases with maturation. In this case, therefore, it would be incorrect to describe material harvested at different stages simply as forage maize; the different maturity stages should be indicated and, on the basis of these, the different types of product classified.

With respect to the processes applied to the feeds, when introducing data on, for example, rice bran meal, it is necessary to know how the meal was obtained and which processes were involved in its production. In the example shown in Table, at least two different types of products (expelled and extracted rice bran meal) need to be differentiated because they are characterized by different chemical compositions and, consequently, by different nutritional values. The process that produces the extracted rice bran meal leads to a material with a much lower fat content compared with the expelled one (7.3 g/kg compared with 90.0 g/kg DM), and this produces a direct effect on the metabolizable energy value. It is apparent that if, instead of specifying the type of process used, the values of the two types of rice bran meal had been averaged, the resulting values would have been meaningless.

Another, more simple, example is related to the on-farm processes applied to forages. Hay is obtained by drying the forage, but the drying process may be achieved by "sun curing" or "barn curing" the material. The different processes can lead to different compositions; for example, related crude protein content varies from 122 g/kg DM for barn-cured hay to 99 g/kg DM for the sun-cured variety. In this case also, information on the type of process applied to the forage is essential in order to maintain the two feeds as distinct entities and, thus, avoid erroneous mean values.

Extrinsic Variability of Data

Differences in analytical procedures may exist among different laboratories and also within the same laboratory. If not identified, differences can compromise the information retrieved from the database as it will compare values provided on different bases. In this respect, an important limitation of databases is related to the type and quality of the data they make available to their users. In general, the diversity of analysis methods adopted by different laboratories is responsible for most of the extrinsic variability found in databases. It is safe to compare or mean parameters only if they have been grouped according to the method of analysis used to obtain them. An example in which incorrect mean values may result is in the estimation of lignin using potassium permanganate, according to the method of Goering and Van Soest, and using acetyl bromide, according to the method of Morrison. These two procedures for measuring lignin not only give different results but are also expressed in different units and it is therefore incorrect, if not impossible, to merge and compare the two values.

Recently, Beever and Mould highlighted large differences in the crude protein and starch contents of maize silages submitted to nine different laboratories. The values observed for one maize silage ranged from 57 to 119 g/kg DM, for crude protein, and from 165 to 272 g/kg DM, for starch. Investigations uncovered the fact that different methods were being used and that some were predicted by near infrared reflectance spectrophotometry (NIRS). Variability also often arises in data on the nutritional value of feedstuffs that have been obtained through *in vivo/in situ* experiments using animals. The animal species may be a source of this inherent data variability. In particular, it is common practice to classify the nutritional value of feedstuffs on the basis of data obtained through measurement among both sheep and cattle under the general definition of "ruminants". Differences in the passage rate of digesta and, therefore, of feed utilization between these two species are often substantial. Other generalizations commonly made arise from the lack of differentiation among the different physiological states of animals of the same species (dry and lactating cows, for example, or growing and finishing pigs). Such physiological differences may have a great impact on the data obtained, and when not recorded will compromise the quality of those data. Moreover, the use of non-standardized feeding and faecal collection protocols can affect the results obtained substantially.

The *in situ* method used extensively to measure the rumen degradability of feeds is subject to many sources of variability. Huntington and Givens reviewed many of these, and the *in situ* method has been the subject of a number of ring tests. For example, Madsen and Hvelplund reported on the protein degradability of five concentrate feeds evaluated *in situ* at 23 different European institutes. Although the mean values obtained were similar to published values, the reproducibility among centres was poor. In particular, the ring test underlined the need to standardize the material used to make the bags and the washing procedures followed.

A last source of variability, in terms of mean data, is related to the number of samples taken of a feedstuff which can vary widely, resulting in consolidated data that are more or less representative. In conclusion, in order to produce a high-quality database on feedstuff composition and nutritional value, the intrinsic variability generally needs to be recorded and differences among materials need to be clearly identified. This is vital if effectively different feeds are not to be merged and averaged but maintained distinct. On the other hand, extrinsic variability should be kept to a minimum because, when a few standard procedures are each applied to a large number of samples, the resulting data will be more representative and useful than when a large number of different methods have each been applied to only a few samples.

EFFECTS OF DATA VARIABILITY FROM FEEDSTUFF

The use of information obtained from feedstuff databases characterized by low-quality data (with high extrinsic variability and low intrinsic variability) may have various effects depending on the end user of the database (*e.g.* animal nutritionist, compound feed manufacturer, farm manager, policy-maker). In this respect, in general, the quality of the data is more often influenced by the extrinsic than the intrinsic variability.

For the Animal Nutritionist

Analysis of feedstuffs composition is costly and time-consuming. Nutritional values acquired through studies with animals are particularly labour-intensive. The nutritionist can therefore not normally afford to analyse specifically each of the feedstuffs involved and often needs to base assessments of nutritive value on more readily available and less expensive information. The animal nutritionist is in charge of designing the animals' diet. The first consequence of unreliable information on feedstuffs, resulting from high extrinsic variability, may be the formulation of a nutritionally unbalanced diet. This may have different effects, depending on nutritional, environmental and economical factors; these effects are highly interrelated since the optimization of feed utilization leads to the optimization of animal output and financial returns.

Nutritional effects. Today, nutritional information on feedstuffs is used to predict animal performance (often with the use of complex models) through estimation of the energy supply to the animal. The incorporation of low-quality data in these models will produce unreliable estimates of animal performance. Furthermore, it is important to maintain the data resulting from *in vivo* measurement of nutritive value separate from those values predicted from chemical composition.

Environmental effects. The most common environmental effect derived from unbalanced diets is caused by nitrogen loss. Particularly for ruminants, the capture of nitrogen depends on the rate of production of microbial protein and is related to the availability of carbohydrates and energy in the rumen. It is therefore important to know the exact amount of degradable nitrogen (together with the carbohydrate and energy availability) when characterizing a feedstuff, in order to capture nitrogen for anabolic purposes and avoid the release of ammonia into the environment.

Table gives an example of how the efficiency of nitrogen conversion in the rumen is affected by the level of nitrogen in the diet. In the Table example, the differences in nitrogen intake are due only to differences in the crude protein content of the grass part of the diet. The balance among nitrogen absorption in the intestine, its conversion into milk protein and its loss in faeces and urine is highly dependent on the capacity of the rumen to capture this nutrient. In the example, increasing the nitrogen intake by approximately 25 per cent resulted in nitrogen losses in the urine increasing by 80 per cent. It is therefore evident that even small differences in the concentration of nitrogen in different feedstuffs may give rise to substantial losses of nitrogen to the environment.

Economic effects. If, for example, maize gluten is fed at the rate of 5 kg per cow per day to a herd of 100 cows at a metabolizable energy (ME) level of 1.8 MJ/kg DM, it would supply the herd with 5 310 MJ ME/day. At a level of 12.8 MJ/kg DM, the daily ME supply to the herd would be 5 760 MJ. Over a 200-day period, the difference in ME supply would be 90 000 MJ, which represents a saving of some 8 tonnes of compound feed, or some US$2 500. Unbalanced diets may also produce an economic loss in terms of animal health, feed conversion efficiency and, ultimately, the output of animal products. The wrong proportion of a nutrient in the diet is reflected in milk production (in terms of both quantity and quality) and has a direct effect on the economics of production. Such effects can be seen extremely rapidly in fast-growing poultry, where even small changes in the nutrient balance can lead to drastic economic consequences.

For the Compound Feed Manufacturer

Compound feeds are composed of a blend of various raw materials and/or by-products. In designing these products, the feed manufacturer needs to find the balanced combination among different materials that will satisfy the requirements of specific animals in a specific physiological state. In order to achieve this, the manufacturer initially needs to base purchasing decisions on the available information about the chemical composition and nutritional value of feedstuffs.

It is therefore very important at this stage that the information available is representative and reliable, not least so that the economic value of the material can be assessed and compared with its market price. Subsequently, the composition of the different ingredients will be measured and the nutritional value of the product estimated, but failure to get the initial values correct will ultimately be reflected in a higher cost of production and will reduce sales of compound feeds to farmers.

For the Farm Manager

The farm manager needs to plan crop production in order to meet the animals' requirements. Managers therefore need to obtain information on the supposed chemical and nutritional characteristics of various ingredients in advance and, based on this information, plan the correct combination and quantity of different forages. If the information is inaccurate, the crops produced will provide unbalanced proportions of nutrients (*i.e.* too much protein, too few available carbohydrates, etc.). The immediate consequences will be an inappropriate diet for the animals, with direct effects on production, and the necessity to compensate the lack of some nutrients by purchasing from the market.

For the Policy-maker

The policy-maker needs reliable information in order to be able to direct the policy of the country or region towards competitive and sustainable agriculture. It is evident that, if estimates are based on imprecise information, policy-makers will not be able to make informed or useful decisions. A further effect of the use of low-quality data by policy-makers is related to the intrinsic variability of feeds, and therefore to real differences in characterizing the materials (genetic differences or differences arising from the processing of raw materials). These differences may be derived from the various ways of describing and naming feedstuffs adopted in different countries. The result is confusion in the trading of feeds, and difficulties in developing and applying export policies.

The lack of harmonization of feed description and naming has been the source of many problems among countries and, in particular, among the Member States of the European Union (EU). Historically, both linguistic and scientific barriers have produced significant differences in feed identification, feed naming and feed description, and false variability often exists because the same name is attributed to different feeds or different names are attributed to identical feeds. The availability of reliable feed data and information that are uniform among different countries will not only contribute to fair trade in feedstuffs but will also facilitate cooperation in animal nutrition research directed towards feed efficiency, product quality and the environment. Not least, it is indispensable for the correct application of the General Agreement on Tariffs and Trade (GATT) and realization of the EU's agricultural policy. Within the EU, a number of steps have been taken to produce a common approach to feed description and naming.

NUTRIENT COMPOSITION OF FEED INGREDIENTS

The feed ingredients potentially available for inclusion in a diet as well as the nutrients they contain must be specified before a least-cost diet can be formulated. Information on available ingredients and their nutrients are traditionally contained in the databases associated with commercial feed formulation packages. This information is frequently updated by the feed formulators.

Decisions need to be made about the actual ingredients to be incorporated in the database and which will be available for consideration during the formulation of any particular diet. The ingredients included will vary between countries and between districts within countries depending on the potential availability of the ingredients for particular livestock enterprises. Traditionally, formulators in the USA have considered mainly corn and soya beans as their primary ingredients, whereas in many European countries and Australia, a wide range of cereal grains, cereal grain by-products, legume seeds, oil seed meals, meat and bone meals, fish meals and manufacturing by-products are included in the databases.

The number of nutrients specified for each ingredient also varies. However, it is essential that the nutrients in the database and their units of expression be the same as those included in the dietary specifications for animal requirements. For example, amino acids for pig diets would be expressed as g of real ileal digestible amino acid/MJ DE to be compatible with the dimensions of the requirements. The database must include also the cost of the ingredient delivered to the site where the feed is made and any additional manufacturing costs associated with particular ingredients. Although most commercial feed formulation packages have the scope to include many hundreds of nutrients, feed formulators often consider a relatively small number, particularly when many diets are being optimised concurrently. However, there are dangers in this practice if for instance only few amino acids are considered during formulation of a diet and some missing amino acids could limit the production of the animals.

MINIMUM AND MAXIMUM INCLUSION LIMITS FOR NUTRIENTS AND INGREDIENTS

Before formulation of a diet takes place, it is important to consider whether limits should be placed on either the minimum or maximum inclusion rates of any nutrients or ingredients. The dietary specification set the minimum limits for most nutrients including amino acids, minerals and vitamins. However, these limits may sometimes be altered based on previous experience of animal performance. In addition, the requirement for each mineral and vitamin is generally not explicitly identified, but many are included in a pre-mix which will be added to the diet at a specified rate. Sometimes specifications are made also for the minimum or maximum available energy content of the diet.

Maximum values are sometimes set for individual nutrients, particularly minerals for which there may be toxicities or imbalances, and for lipid and fibre because of their ability to interfere with the digestion of other dietary components. Minimum and maximum limits are set also to ensure that the ratio of interacting nutrients such as calcium to phosphorus or of unsaturated to saturated fatty acids is within a desired range. Furthermore, restrictions may be imposed on the amounts of total nitrogen and phosphorus that will be included in a diet to limit the polluting potential of the effluent. Maximum limits are often placed on the inclusion rate of specific ingredients for reasons such as palatability, potential for toxicity, limits to digestive capacity of the animal and potential incorporation of excessive moulds, mycotoxins, weed seed toxins or hazardous chemical residues.

Minimum limits are often imposed to force the inclusion of ingredients that are considered to be desirable because of either their high palatability or perceived growth promoting effects, such as milk powders and porcine plasma which stimulates antibody production. Maximum or minimum inclusion rates of ingredients are imposed also to ensure that the final product has a suitable physical structure to improve the ease of flow through the mill and at the farm reduce dustiness or improve the integrity of pellets.

MANUFACTURE OF THE FEED

The exercise of delivering specified nutrients to the animal in question is not completed until the animal has consumed an appropriate amount of feed, which contains the nutrients intended. If the manufactured feed either bears little resemblance to the theoretical formulation or contains toxic and unpalatable materials, the performance of the stock will be compromised and the effort involved in the accurate formulation of the feed wasted. Three issues require special attention during feed manufacture. First, the chemical composition of each ingredient actually used in the feed mix should be known. Most commercial feed mills now use near infrared spectroscopy to determine the moisture, protein, lipid and fibre contents of ingredients delivered and this information is used to specify the composition of the ingredients considered in the least-cost formulation process.

In addition, ingredients should be screened for contaminants such as moulds, mycotoxins and toxic weed seeds as well as for rancidity and other factors likely to affect the palatability of the diet, feed intake or animal performance. Secondly, the ingredients must be accurately weighed and completely mixed. Numerous factors such as overfilling of mixers, worn equipment, build up of ingredient residues, inappropriate mixing times and post-mixing segregation can lead to wide variation in the uniformity of the final product.

Thirdly, the finished product should be in a form that is efficiently utilised by the animals and that flows readily through the equipment at the mill and farm, and is not excessively dusty. It is important the physical nature of feed on wastage. Pigs weighing 25 kg in groups of 10 were offered ad libitum for four weeks the same feed either as a meal or in pelleted form. Feed intake and growth rate were unaffected by the physical form of the feed, but feed waste was many times greater for pigs consuming the meal.

Table. Effect of the Form of Feed on the Fed Wastage and Performance of Pigs

	Meal	Pellets
Feed intake (g/day)	1551	1640
Gain (g/day)	771	789
Feed conversion (g/g)	2.48	2.13
Wastage (g/day)	388	26
Wastage (per cent)	18.7	1.5

The most important factor determining the success of the feed manufacturing process is the need for vigilant and rapid quality control capacity. It is essential that critical control point analyses are conducted on the manufacturing operation and that standard operating procedures are documented and followed. However, the translation of these requirements into dietary specifications still requires considerable judgement and experience. Judgements are required about the actual factors determining the performance of stock under commercial conditions, their likely feed intake, the units of expression of the nutrient requirements and the actual nutrients to be included in a formulation.

Considerable effort has been put into incorporating ingredients into databases, which are normally associated with the least-cost formulation software packages. However, judgements are also required in the setting of minimum and maximum inclusion rates for both nutrients and ingredients. The actual formulation of the diets is straightforward, but it is important to determine whether the feed will be optimised on a cost/tonne basis or a cost/MJ energy basis. The final manufacture of the feed requires considerable attention to detail and quality control. Many of the steps in diet formulation have been made more accurate and less tedious since the advent of accurate computerised prediction models.

TRANSLATION OF TISSUE REQUIREMENTS INTO DIETARY SPECIFICATIONS

Translation of animal tissue requirements into dietary specifications has traditionally been one of the most difficult tasks in diet formulation because numerous judgements must be made about the actual production characteristics and housing environment of the animals under consideration and about the units in which the requirements should be expressed. Judgements need to be made about the genotype and food intake of the animals, whether their performance and, therefore, their nutrient requirements are affected by climate, activity, disease, penning environment and other factors. In most of the earlier systems that relied on the tabulation of dietary requirements, the effects of the animal's genetics and of environmental factors were generally ignored. However, now with computer simulation models being used frequently to determine dietary specifications, the interactions between the animal and its environment on dietary specifications can be readily determined.

Once the nutrient requirements for the animals in question have been determined, the second major decision is about the units in which these requirements will be expressed. For example, amino acid requirements could be expressed as amounts available at the tissues, truly digested, apparently digested in the small intestines or apparently digested over the whole digestive tract. Once this decision is made, it is necessary to determine whether the amino acid requirements are to be expressed as g/day, g/unit of energy or g/kg diet. Further decisions are needed to determine whether specifications will be made for all essential amino acids and whether they will be expressed in the selected units or as a ratio to lysine. If the amino acids are to be expressed relative to energy, should it be gross energy (GE), digestible energy (DE), metabolisable energy (ME) or net energy (NE)?

Then, once all these decisions have been made and the units defined, a final decision is required about the live weight range or physiological state over which the specifications should apply; that is, how many diets should be fed to animals as they grow from weaning to slaughter; will the specifications for these diets be taken from the beginning of the weight range, the middle or at some other point? The most accurate dietary specifications are for amino acids available at the tissues in relation to NE. However, the dietary specifications for the animal must be in the same units as the information on the feed ingredients that could be considered during diet formulation. There is very little information currently on the true availability of amino acids in feedstuffs, whereas there is a reasonable amount on the true ileal digestibility and even more on apparent ileal digestibility.

Because endogenous amino acid losses from the gut are significantly affected by feed intake, the fibre, protein and antinutritional content of the diet and the bodyweight of the animal, the best and most practical units for amino acids appear to be g real ileal digestible amino acid per unit of energy. NE is a function of both the feed and the animal and is not constant for an ingredient. However, as there is little difference in the ratio between the ME and DE contents of most ingredients and extensive information is available for the DE content of diets for pigs and apparent metabolisable energy (AME) contents of diets for poultry, the most practical units for specifying dietary amino acids appear to be as g real ileal digestible amino acid/MJ DE in pigs and/MJ AME in poultry.

Ideally, each essential amino acid plus the combinations of methionine plus cystine and of phenylalanine plus tyrosine should be specified in this way.

However, many feed formulations are based on lysine expressed as g real ileal digestible lysine/MJ DE with other essential amino acids being expressed as a ratio to lysine. Many feed formulators over specify lysine as an insurance against deficiency and therefore over specify the other essential amino acids considered in the formulation. Current sources of dietary specifications are tables such as ARC, NRC and SCA, results from specific trials where the response of animals to increasing dietary concentrations of a nutrient has been examined, experience of feed formulators and animal feeders or computer simulation models like AUSPIG and the new NRC calculations.

FORMULATION OF DIETS TO LEAST-COST

The process of least-cost formulation of feeds using linear programming methods is relatively straightforward and has been well defined in commercial software packages available to industry. The procedure is to match the dietary nutrient specifications for the animal with the lowest cost supply of nutrients from potentially available ingredients within the minimum and maximum limits set for specific ingredients. Fundamental to the process is that the nutrient requirements for the stock and the nutrient values of the feed ingredients are in the same units.

A significant assumption in the linear programming procedure is that there is no interaction between ingredients in the availability of nutrients. That is, the nutrient contents in all ingredients are assumed to be additive in the formulated feed. Although this assumption is reasonable for most situations, it is not true when specific ingredients which contain large amounts of lipid, fibre or antinutritional factors (like tannins) are included in the feed. A common approach to least-cost feed formulation is to minimise the cost per unit weight of feed. The difficulty with this method is that the quantity of nutrients in a fixed weight of feed can vary widely and may not provide nutrients at least-cost. In pig diets, DE provides an accurate measure of the total nutrient content of the diet because amino acids and other nutrients are specified in relation to DE.

The cost of diets formulated to least-cost/tonne increased slightly as the DE content of the diet was raised from 12 to 13 MJ DE/kg and then rapidly with further increases in the DE content of the diet to 16 MJ DE/kg. However, the cost of these diets on a cents/MJ DE basis fell steeply from the 12 MJ DE/kg diet to a minimum at 13.5 MJ DE/kg and then rose steeply again to 16.0 MJ DE/kg (Fig. 8.). If diets were formulated on a weight basis without a DE restriction, the 12 MJ DE/kg diet would be chosen, but clearly at a higher cost per nutrient. However, if the diets were formulated on an available energy basis, the 13.5 MJ DE/kg diet would be chosen. The latter method of formulation supplied nutrients at the lower cost.

5

Proteins, Amino Acids and Peptides in Animals

Proteins are complex, organic compounds composed of many amino acids linked together through peptide bonds and cross-linked between chains by sulfhydryl bonds, hydrogen bonds and van der Waals forces.

There is a greater diversity of chemical composition in proteins than in any other group of biologically active compounds. The proteins in the various animal and plant cells confer on these tissues their biological specificity.

Proteins can be classified as:

- Simple proteins. On hydrolysis they yield only the amino acids and occasional small carbohydrate compounds. *Examples are*: albumins, globulins, glutelins, albuminoids, histones and protamines.
- Conjugated proteins. These are simple proteins combined with some non-protein material in the body.
  Examples are: nucleoproteins, glycoproteins, phosphoproteins, haemoglobins and lecithoproteins.
- Derived proteins. These are proteins derived from simple or conjugated proteins by physical or chemical means.
  Examples are: denatured proteins and peptides.

The potential configuration of protein molecules is so complex that many types of protein molecules can be constructed and are found in biological materials with different physical characteristics. Globular proteins are found in blood and tissue fluids in amorphous globular form with very thin or non-existent membranes. Collagenous proteins are found in connective tissue such as skin or cell membranes. Fibrous proteins are found in hair, muscle and connective tissue. Crystalline proteins are exemplified by the lens of the eye and similar tissues. Enzymes are proteins with specific chemical functions and mediate most of the physiological processes of life. Several small polypeptides act as hormones in tissue systems controlling different chemical or physiological processes. Muscle protein is made of several forms of polypeptides that allow muscular contraction and relaxation for physical movement.

PROPERTIES

Proteins can also be characterized by their chemical reactions. Most proteins are soluble in water, in alcohol, in dilute base or in various concentrations of salt solutions. Proteins have the characteristic coiled structure which is determined by the sequence of amino acids in the primary polypeptide chain and the stereo configuration of the radical groups attached to the alpha carbon of each amino acid.

Proteins are heat labile exhibiting various degrees of lability depending upon type of protein, solution and temperature profile. Proteins can be reversible or irreversible, denatured by heating, by salt concentration, by freezing, by ultrasonic stress or by aging. Proteins undergo characteristic bonding with other proteins in the so-called plastein reaction and will combine with free aldyhyde and hydroxy groups of carbohydrates to form Maillard type compounds.

CHEMICAL DETERMINATION

The nitrogen content of most proteins found in animal, nut and grain tissue is about 16 per cent; therefore, protein content is commonly expressed as nitrogen content × 6.25.

PROTEIN DIGESTION AND METABOLISM

Ingested proteins are first split into smaller fragments by pepsin in the stomach or by trypsin or chymotrypsin from the pancreas. These peptides are then further reduced by the action of carboxypeptidase which hydrolyses off one amino acid at a time beginning at the free carboxyl end of the molecule or by aminopeptidase which splits off one amino acid at a time beginning at the free amino end of the polypeptide chain.

The free amino acids released into the digestive system are then absorbed through the walls of the gastro intestinal tract into the blood stream where they are then resynthesized into new tissue proteins or are catabolysed for energy or for fragments for further tissue metabolism.

ADVANCES IN PROTEIN TECHNOLOGY

DNA micro arrays have become ubiquitous in genomic research, evident by their widespread use in profiling gene expression patterns, mapping novel transcripts, detecting sequence mutations and deletions, and locating transcription factor binding sites. Although micro array experiments are invaluable for large-scale sequence analyses, little can be inferred from these studies about the functions of gene products.

In contrast to the high-throughput (HT) experiments afforded by DNA arrays, those designed to elucidate the biochemical activities of encoded proteins have traditionally been carried out on single molecules. Recently, significant effort has been made towards adapting proteomic analytical methods for use with DNA micro array technologies to enable the elucidation of proteome-wide biochemical activities and interactions.

The large-scale characterization of protein complexes generally involves:

- The separation of complex protein samples and
- The subsequent identification of individual proteins.

Among the methods currently available for proteome analysis are 1D gel electrophoresis (1D-GE) and 2D gel electrophoresis (2D-GE), MS, affinity chromatography, N-terminal Edman protein sequencing, metal affinity shift assays, N isotope labeling, and tandem affinity purification (TAP) tagging to isolate protein complexes from cell extracts. Of these, protein separation by 2D-GE and subsequent identification using MS have remained the two core technologies for large-scale proteomics. The 2D-GE method entails the separation of complex protein mixtures by molecular charge in the first dimension and by mass in the second dimension.

Although recent advances in 2D-GE have improved the resolution and reproducibility, the technique remains difficult to automate in a HT setting. For this reason, alternative approaches that obviate the need for gel separation, such as multidimensional protein identification technology, have gained popularity for large-scale proteomics efforts and are able to generate a comprehensive catalogue of proteins present in complex cell extracts. HT protein analysis is expected to accelerate with the introduction of new robotic liquid chromatography systems and high-resolution analysis methods such as top-down Fourier transform mass spectrometry.

The use of MS for protein identification has come into wider use with the advent of soft ionization techniques, such as electrospray ionization (ESI) and MALDI. Additionally, the emergence of hybrid methods incorporating electrospray technology with quadrupole time-of-flight mass spectrometer tandem mass analysis (ESI Q-TOF MSD MS) allows more accurate identification of specific proteins through the generation of collision-induced dissociation (CID) spectra that yield accurate sequence tags from protonated peptide ions.

Recently, the range of mass spectrometric applications has been extended by other tandem approaches such as MALDI TOF–TOF MSDMS and MALDI Q-TOF MSDMS. Beyond the identification of individual proteins, quantitative analysis of complex samples can be accomplished through the use of surface-enhanced laser desorptionDionization (SELDI). This approach incorporates standard ionization techniques on different surfaces, comprising a solid support modified with various chemical or biological bait molecules. These may include antibodies, proteins, nucleic acids and metal ions.

The differential surface capture of solubilized protein samples provides a unique signature that varies depending on protein composition. Unlike MS techniques, SELDI is not able to identify specific proteins in a complex sample. This is expected to change in the near future through the combined use of

SELDI technology with tandem mass spectrometers. Regardless of the methods used to measure and catalogue an organism's proteome, the majority of detection and quantification methods result in denaturing of the protein samples and thus functional characterization is not possible.

To obtain detailed functional information, proteins must be cloned and expressed in recombinant form and subjected to systematic biochemical analyses. Martzen *et al.* developed a multiplexed assay to characterize protein function on a large scale through a divide-and-conquer strategy. The approach entails the generation of pooled purified protein samples, which are then assayed in parallel for various biochemical activities. Individual proteins that exhibit specific activities in a pooled sample are identified through a series of recurrent analyses of sub pools. This procedure allows the rapid identification of proteins that participate in various biochemical pathways via a divisive search through subpopulations of functional proteins.

An advantage of this method is that biochemically active, multimeric protein complexes may be identified in vitro; however, this does not represent an exhaustive combinatorial search as only proteins that happen to be present in a given pool may interact.

DEVELOPMENT OF ADDRESSABLE PROTEIN ARRAYS

The ultimate goal of protein micro array development is to construct ordered arrays of individual proteins to assess biochemical activities on a single-molecule basis. One solution has entailed the use of analytical arrays for the purpose of protein profiling. These typically comprise a library of peptides or antibodies arrayed on the surface of a glass microscope slide. In general, protein profiling entails the measurement of binding specificity, affinity, or abundance of proteins in a biological sample. To address this with micro arrays involves the construction of an array whose elements are designed to capture, and thereby measure the binding specificity of, proteins present in a complex mixture such as a cell lysate or serum sample.

Array features are typically antibodies that are mechanically printed as independent purified samples, but may also be peptides that are synthesized in situ. The latter technology holds promise for the rapid screening of high-affinity binding sequences and the identification of potential drug targets. The synthesis of individual peptides in situ can be accomplished via photolithography, originally described by Fodour *et al.* and later applied to oligonucleotide-based micro array fabrication.

Photolithographic peptide arrays involve the use of photolabile phosphoramitides that enable deprotection of the Boc group from an amino moiety, allowing polymer synthesis to proceed at discrete array locations when illuminated by a laser. Using this method, an array of 1024 peptides was constructed and probed with a mAb. Pellois *et al.* developed an alternative technique that can make use of natural amino acids for peptide synthesis, using a photogenerated acid to chemically deprotect the growing polymer chain upon exposure to light.

A simpler approach to photolithography is described by the SPOT protocol, in which a series of activated amino acids is mechanically deposited onto a porous surface, thereby building the desired peptides sequentially. SPOT is based on conventional solidphase synthesis chemistry, and may therefore be more accessible in terms of implementation. Recently, Li *et al.* described a novel approach to homogeneous in situ peptide synthesis based on a common cyclic peptide scaffold.

The procedure involves the deprotection of NPPOC phosphoramitide groups to affect the addition of side-chains to a universal core molecule, which is presythesized and applied to a silanized glass slide in a uniform manner. A library of individual peptides can then be synthesized in situ using maskless photolithography, in which a spatially addressable array is fabricated through successive photodeprotection using a bank of digitally controlled micromirrors. To date, most protein micro array systems have been based on contact-printed antibody libraries that are used for profiling complex analyte mixtures.

The most widely adopted strategy consists of a multiplex adaptation of the classical antibody sandwich assay, where a pair of antibodies binds two discrete recognition surfaces on each protein. In this procedure, an array of antibodies, which have been immobilized through covalent bonding to a silanized glass surface, is probed with various analytes. A second biotinylated antibody is then applied which binds to captured analytes, forming an immune complex.

The second antibody is finally detected with a universal antibiotin antibody conjugated to a fluorophore. This approach has been used to great effect for the simultaneous detection of multiple cytokine or chemokine levels in biological samples. The highly specific antibody–complex recognition is ideal for detecting low-abundance cytokines and holds much potential for clinical diagnostic applications and discovery of therapeutic drug targets. A variation on this experiment utilizes rolling circle amplification (RCA) to enhance the fluorescence signals emitted from the immune complex.

In this method, the detection antibody is not fluorescence-labeled but is instead conjugated to oligonucleotides that serve to prime the RCA reaction. Complementary circular oligonucleotides are extended with DNA polymerase, producing RCA products that consist of tandem repeats. Because these repeat sequences provide many redundant hybridization targets, an amplification in fluorescence signal is achieved when the RCA products are detected with fluorescence-labeled complementary DNA probes. An impediment to the further development of antibody array technology lies in the availability of highquality antibodies against the individual proteins in a complex sample.

At present it remains unfeasible to obtain hundreds or thousands of different antibodies that can recognize and capture various proteins with high affinity and specificity–factors that are essential for preventing cross-reactivity. The problem is compounded when considering multiplex sandwich immunoassays, where two highly specific antibodies must be obtained for each protein captured. These must also recognize two different regions of the protein, each without masking the other binding domain. Although this issue may eventually be addressed by new methods of HT antibody generation, several alternatives to protein capture have been developed that do not rely on antibody recognition.

Among the most innovative of these involves imprinting technology to create artificial molecular recognition surfaces. Peptides that correspond to signal sequences in various target proteins are used as a structural scaffold, around which polymerizable monomers are allowed to self-assemble. The monomers are crosslinked in place and the template molecule is stably removed from the complex. The cavity or imprint that remains is shapecomplementary to the original template and will therefore bind identical structures with high affinity.

This technology promises to accelerate antibody array development by increasing the throughput of artificial epitope production, although at present imprinting is unable to mimic larger functional proteins or other macromolecular structures. The antibody array represents an excellent platform for HT protein profiling. However, the large-scale study of protein biochemistry using the micro array format requires the development of arrays of full-length, functional proteins.

HT protein production, combined with technologies shared with the proven DNA micro array format, allow the simultaneous analysis of thousands of protein activities in a single experiment. In addition to enabling the inverse profiling experiment where functional proteins can be interrogated with individual antibodies, protein micro arrays enable a wide range of biochemical assays in response to any solution-phase binder.

ARRAY TECHNOLOGIES FOR FUNCTIONAL PROTEIN ANALYSIS

The principal challenges in functional protein array development comprise:

- Creation of a comprehensive expression clone library,
- HT protein production, including expression, isolation and purification,
- Adaptation of DNA micro array technology to accommodate protein substrates,
- Ensuring the stability of arrayed proteins, and
- Reduction of inter-and intraslide variability of protein concentration between deposited samples.

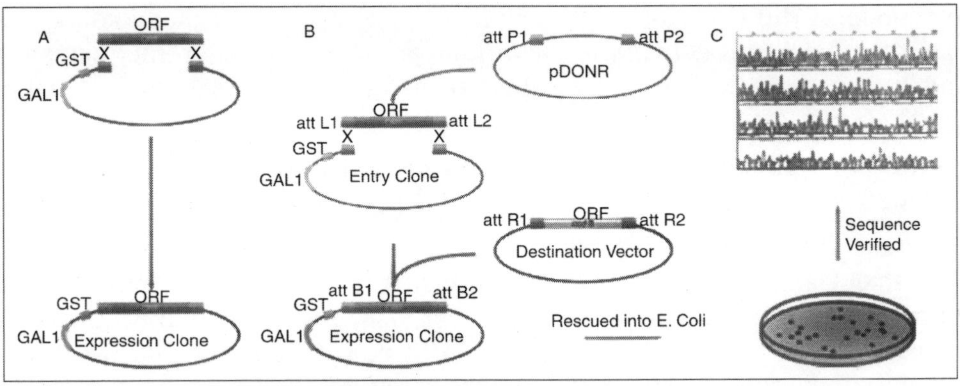

Fig. Example of High-throughput Cloning Strategies for Expression Library Construction.

(A) The construction of in-frame gene fusions used by Zhu *et al.* relies on gap repair-mediated recombination in yeast. Amplified ORFs are mixed with a linearized vector with ends that are identical to those of the amplified DNA, and the mixture is transformed into yeast cells where the DNA is integrated.

(B) Recombination cloning based on the Gateway k-phage integration D excision system (Invitrogen, Carlsbad, CA, USA). Recombination between the k att P and host att B sites yields an integrated phage with ends att L and att R. During excision, recombination between att L and att R sites regenerates att P and att B. Amplified ORFs with att B1 and att B2 sites undergo recombination with a donor vector containing att P1 and att P2 sites, producing the entry vector with att L1 and att L2 sites. Subsequent recombination with a destination vector containing att R1 and att R2 sites allows transfer of the ORF into the expression vector, regenerating att B sites.

(C) Expression clones are rescued into Escherichia coli and verified by DNA sequencing.

By far the greatest obstacle in developing functional protein micro arrays is the construction of a comprehensive expression clone library from which a large number of distinct protein samples can be produced. In building a clone library, it is desirable to construct recombinant genes where fusion proteins can be produced for the purpose of affinity purification and D or slide surface attachment.

Cloning the genes of interest with an inducible promoter allows individual proteins to be expressed in high abundance. HT purification can be accomplished with the addition of C-or N-terminal tags, such as glutathione-S-transferase or the IgG-binding domain of Protein A.

The incorporation of fusion tags also facilitates the verification of clone inserts by sequencing across the vector–insert junction. It is highly desirable to transform the expression vector into a homologous or related cell type, ensuring the proper delivery of the protein product to the secretory pathway and hence correct folding and post-translational modification of each recombinant protein.

PROTOTYPE FORMATS FOR FUNCTIONAL PROTEIN ARRAYS

A critical aspect of the development of arrays of functional proteins is the selection of an experimental support and its associated method of surface attachment and immobilization of proteins. If proteins can be immobilized without disrupting their native conformations, they are likely to remain biologically active in vitro. The method of immobilization will also greatly influence the orientation in which proteins are attached to the support surface.

Engineering a common point of attachment for all samples, typically through the inclusion of affinity tags, ensures that at least a subset of proteins maintain a uniform presentation to solution-phase binders. The array format must be compatible with appropriate detection instruments and exhibit a wide dynamic range of intensity values associated with protein binding or catalysis events. Additionally, non-specific binding of labeled samples should be minimized in order to reduce the background and increase the signal-to-noise ratio of the experimental platform.

The materials traditionally associated with conventional protein assays are often incompatible with robotic arrayers, cannot provide the sensitivity or dynamic range expected from micro array experiments, or contribute high fluorescence background resulting in low signal-to-noise ratios. To circumvent these problems, a number of innovative platforms have been explored for prototyping the protein array format.

Among these, two make use of an agarose or acrylamide gel situated on glass slides, thereby combining the utility of a solid support with the loading and binding capacity of a porous gel matrix. These methods were originally devised to increase the potential loading capacity of DNA samples on a planar micro array surface, but are generally applicable to a variety of substrates, including nucleic acids, proteins and small molecules. Guschin *et al.* developed arrays of polyacrylamide gel pads on a hydrophobic glass surface using a combination of gel photopolymerization and manual contact pin deposition. Three test proteins were loaded onto the arrays–mouse IgG1, rabbit IgG and BSA.

Immunoanalysis of fluorescein isothiocyanate (FITC)-labeled IgG against mouse IgG1 demonstrated selective binding when the arrays were imaged with a fluorescence microscope. In a related study, Mirzabekov and coworkers experimented with a copolymerized acrylamide–bisacrylamide substrate, producing arrays of discrete gel pads of between 10 and 100 lm in diameter. In a later study, Kiyonaka *et al.* developed a method of supramolecular gel formation as a spontaneous process that does not require additional polymerization steps, using the approach to develop a sensor array of fluorescent metal anion and cation receptors in a glycosylated amino acetate hydrogel matrix.

A study by Afanassiev *et al.* explored the use of a thin, uniform layer of agarose film on a glass surface to achieve a similar effect. Activated agarose containing NaIO was applied to silanized slides, and samples were mechanically deposited using a robotic microspotter after the gel had solidified. A limited application of this platform to protein-based assays demonstrated the binding of mAbs to immobilized recombinant human BAD protein, and reciprocally, of recombinant human (rh)BAD to immobilized antibodies in a sandwich immunoassay.

Another approach involved the use of a liquid silica compound to create flexible sheets of microwell arrays in which biochemical reactions are performed en masse. Zhu *et al.* devised a system of casting a silicone elastomer (polydimethylsiloxane) onto a reusable mold of laser-milled acrylic. After the microwell sheets had cured, various molecules were immobilized to the interior surface of the wells using the chemical crosslinking agent, 3-glycidoxypropyltrimethoxysilane.

Aside from demonstrating the technical feasibility of protein immobilization and binding to various molecules in vitro, it is essential to conduct biologically relevent experiments if protein micro arrays are to become an established research platform. To explore the utility of microwell arrays for the detection of enzymatic activities, Zhu *et al.* focused on 119 protein kinases from the budding yeast Saccharomyces cerevisiae (95 known and 24 uncharacterized).

Each of the protein kinases was expressed in recombinant form and purified, then assayed for the ability to phosphorylate 17 substrate proteins. A total of 17 arrays were fabricated and one of the substrate proteins was immobilized in every well; a different protein kinase was delivered into each well in the presence of γ-ATP to determine which kinases were capable of phosphorylating the given substrate protein. The arrays were incubated and then washed such that all free enzyme was removed, and the signal from the radiolabeled proteins in each well were quantified using a phosphoimager.

The possibility that measurements originated from the autophosphorylation of the kinases themselves was discounted, as in each case the substrates were bound in the wells while the enzymes remained free in solution; the wells were cleared of this reaction mixture and washed prior to imaging.

In addition to detecting expected phosphorylation activities, 27 protein kinases were found to be capable of phosphorylating tyrosine after incubation with a synthesized poly(tyrosine-glutamate) peptide. This finding was significant as yeast protein kinases are generally known to phosphorylate only serine or threonine.

Phylogenetic analysis revealed sequence similarities among the tyrosine-phosphorylating kinases that were specific to several amino acids oriented in the catalytic cleft and substrate-binding domain of the enzyme. These appeared exclusively in the kinases that phosphorylated tyrosine and not in those that phosphorylated serine or threonine alone.

Although it is likely that most yeast protein kinases will preferentially phosphorylate serine or threonine in vivo, this study demonstrated that protein arrays are sensitive enough to reveal previously uncharacterized biochemical properties in a HT assay.

On balance, each of these formats retains a number of important properties for proteomic experiments. Microwells provide the ability to preserve native protein function by carrying out reactions in an aqueous environment, while the hydrated gel matrix approach also immobilizes the proteins to some degree. Any potential cross-contamination between arrayed proteins is eliminated owing to physical barriers between them or the spatial separation of discrete gel pads.

Each platform affords a far greater loading capacity than the planar surface of a glass slide, allowing experiments to be performed at varying substrate concentrations. Finally, the microwell format offers the potential to perform complex, multistage experiments in solution if reaction mixtures are exchanged using microfluidic robotics.

CONTACT-PRINTED FUNCTIONAL PROTEIN MICRO ARRAYS

Although these approaches offer numerous technical advantages, increasing attention has been paid to adapting existing glass-slide micro array technologies for use with proteins. As a result, the practice of printing directly onto chemically treated glass surfaces, apart from their indirect function as a solid support, is now in wider use for protein arrays.

The principal motivating factor for using glass slides is to take advantage of robotic microspotting arrayers and laser scanners that have become commonplace for DNA micro array fabrication and image acquisition, respectively. The standard micro array format therefore affords the opportunity for a broader range of investigators to adopt HT proteomic formats using existing DNA micro array construction and analysis equipment.

Table. Surface Chemistries for Glass Slide Protein Micro Arrays

Surface Chemistry	Protein Attachment	Protein Orientation	Modifications Required
Epoxy	Covalent crosslinking	Random	None
Aldehyde	Covalent crosslinking	Random	None
Poly(L-lysine)	Adsorption	Random	None
Nitrocellulose	Adsorption, absorption	Random	None
Poly(vinylidene difluoride)	Adsorption, absorption	Random	None
Avidin	Affinity binding	Random	Biotinylation
Nickel-nitrilotriacetic acid	Affinity binding	Uniform	His6 fusion

A number of different chemical treatments are suitable for the immobilization of proteins on glass, including poly(l-lysine), aldehyde, nickel, epoxy, avidin and nitrocellulose. The choice of surface chemistry will determine the method of protein attachment to the functionalized support, which in turn will influence whether the proteins require modification prior to arraying.

Amine-reactive coatings (*e.g.* 3-aminopropyltriethoxysilane) do not require proteins to be modified for effective immobilization, as covalent attachment is achieved through the random crosslinking of amine groups. Protein attachment to either nickel-or avidin-coated slides is made through affinity binding to histidine residues or biotin, respectively, and therefore requires a recombinant approach to construct fusion proteins.

The advantage of non-covalent immobilization is that because attachment takes place exclusively at the affinity tag, a significant subset of proteins will be immobilized in a presumably uniform orientation where their functional domains are exposed to solution and available to interact with the labeled sample.

Other membrane-based solutions, such as fluorescent array surface technology slides, facilitate protein immobilization through passive adsorption on a nitrocellulose surface, and in many cases yield superior signal-to-noise ratios in comparison with poly(l-lysine) or aldehyde-treated slides.

- (A) Control proteins are spotted in a dilution series and illuminated with fluoresceinlabeled antibodies.
- (B) Protein–protein interaction assays reveal specific binding targets.
- (C) Detection of lectin modifications across the yeast proteome.

ROBOTIC PRINTING OF PROTEIN MICRO ARRAYS

Constructing micro arrays from purified proteins is more challenging than building their DNA counterparts. Aside from the complexities of preparing hundreds or thousands of individual protein samples, the proteins must remain active during the manufacture and probing of micro arrays. This entails keeping protein samples in cold storage for as long as possible, and arraying them in a glycerol solution to maintain their native state on the slide surface.

The procedures used for printing protein arrays are similar to those developed for DNA arrays, but with several important differences. DNA samples are arrayed in a high-humidity chamber by a robotic microspotter. Although the same equipment can be used to print protein arrays, individual samples are typically handled in a 30–35 per cent glycerol solution to prevent denaturing, and arrayed at concentrations in the range of 0.3–1 mgÆmL.

As glycerol is hygroscopic, care must be taken to reduce the ambient humidity to 28–30 per cent; this differs from the 45 per cent humidity environment in which DNA arrays are usually printed. Printed spots of protein–glycerol solution will absorb moisture from the air in high-humidity environments and begin to expand on the slide surface. This can lead to cross-contamination of samples if the proteins are arrayed at high spatial density.

The diameter and pitch (the spacing between samples) of spots can remain similar to the parameters used for printing DNA micro arrays, although this will depend on the type of slides used and the viscosity of the buffer. For example, epoxy-coated slides provide more efficient protein attachment, which generally allows for higher-density printing. Pilot studies should be carried out to determine the appropriate printing density for a given sample preparation and surface chemistry, adjusting the pitch from a more conservative spacing of 1000 μm down to a minimum of 300 lm. Depending on the number of pins used (typically between 16 and 48), a 300 μm pitch allows up to 20 000 individual protein samples to be printed on a standard microscope slide, yielding ≈1600 features per cm2 where each feature is 150 μm in diameter.

A and *C*—Protein–protein interactions were detected with fluorescein-labeled antibodies to specific proteins. In this case, BODIPYFL–IgG, Cy3–IjBa and Cy5–FKBP12 identify the presence of protein G, p50 and FKBP–rapamycin-associated protein, respectively. *B*—Phosphoinositide-binding specificity to functional proteins is determined in a phospholipid probing assay.

The detection of protein–protein and protein–lipid binding involves a two-step process. First, the sample used to probe the array is biotinylated using a simple crosslinking method. To perform the experiment, the micro array is incubated with a solution containing the biotinylated sample, washed, then incubated a second time in the presence of a fluorescent streptavidin molecule.

The actual protein interactions are formed during the first incubation; subsequent binding of the streptavidin–fluor conjugate to the biotin label then allows these interactions to be detected using a micro array scanner. A simpler approach can also be employed by attaching a fluorophore directly to the solution-phase binder used to probe the array. Proteins can be labeled using amine-reactive dyes, where the succinimidyl ester moieties react with the primary amines of the protein to form stable conjugates.

As with DNA arrays, fabrication precision should be assessed relative to the printing consistency across all of the features on the array, consistency between features from multiple arrays, and the uniformity of sample concentrations. This is typically measured as the coefficient of variation (CV), which estimates the reproducibility of printing as the standard deviation over the mean fluorescence signal intensity.

For functional protein arrays prepared from individual purified samples, intra-array CV measurements fall in the range of 15–20 per cent, and are a measurement of both the variability in protein concentrations and the pin-to-pin consistency of sample deposition during the printing cycle. Inter-array CVs over all protein samples from replicate slides are typically lower, in the range of 12–15 per cent. Much of the variation in sample concentration is attributed to the expression and purification of lowabundance proteins, whereas printing variation can arise as a function of the order in which samples are arrayed during long print runs. The incorporation of an epitope tag, such as glutathione-S-transferase, is invaluable for HT affinity purification, and also allows the global quantification of protein signals by probing the array with a fluor-conjugated antibody.

This measurement is used to calculate the ratio of the probe-binding signal relative to the baseline fluorescence intensity of each spot on the array. As the relative concentrations and deposited amounts of protein samples will vary slightly, computing the fluorescence values in terms of intensity ratios enables the application of statistical methods similar to those used for DNA micro array data analysis.

DETECTING BIOCHEMICAL INTERACTIONS

Depending on the type of assay being performed, protein–substrate binding or catalysis can be detected via fluorescence labeling, chemiluminescence, radioisotope labeling, or label-free methods, such as surface plasmon resonance imaging, atomic force microscopy or reflectometric interference spectroscopy. Fluorescence labeling is generally preferred as a safe and efficient method for HT analysis that is compatible with current micro array laser scanners.

The detection of protein binding, as described in Zhu *et al.*, involves a two-step process similar to that used for multiplex sandwich immunoassays. First, the sample used to probe the array is biotinylated using a simple crosslinking method. To perform the experiment, the micro array is incubated with a solution containing the biotinylated sample, washed, then incubated a second time in the presence of a fluorescent streptavidin molecule.

The actual protein interactions are formed during the first incubation; subsequent binding of the streptavidin–fluor conjugate to the biotin label then allows specific interactions to be detected using a standard micro array scanner. A simpler approach can be employed by attaching a fluorophore directly to the solution-phase binder used to probe the array, provided that the label does not mask an active site or binding domain of the sample molecule. Proteins can be labeled directly using aminereactive dyes, where the succinimidyl ester moieties react with the primary amines of the protein to form stable conjugates. The labeled protein can then be purified using size-exclusion spin columns and applied to the array for the detection of protein–protein interactions. Of course, other molecules can be applied to functional protein arrays to assess different molecular interactions, such as protein–DNA binding. Fluorescence labeling of DNA can be achieved in a variety of ways, such as through direct incorporation of fluorescent-labeled nucleotides during reverse transcription, or by secondary conjugation of amine-reactive succinimidyl esters to 5-(3-aminoallyl)-DUTPs.

FUNCTIONAL PROTEIN ARRAYS IN USE

The first study reporting the use of contact-printed, glass-slide protein arrays was described by MacBeath and Schreiber, who investigated the binding activities of three known pairs of interacting proteins. One protein of each pair was printed in quadruplicate onto aldehyde slides, and the arrays were probed with the labeled partners. The group also explored various feature densities of printed protein samples, successfully arraying a single protein as 10 800 discrete features on a standard micro array slide.

Importantly, the researchers were able to quantify the concentrations of the bound and solutionphase proteins necessary to carry out the experiments. Sample concentrations between 100 μg. mL^{-1} and 1 mgÆ mL^{-1} were found to be suitable for protein immobilization and detection, whereas solution-phase proteins at concentrations of ≈12.5 pm yielded fluorescence signals that scaled linearly over four orders of magnitude. Thus, these experiments demonstrated the feasibility of arraying proteins in a standard micro array format and at feature densities comparable with those of contactprinted DNA arrays.

A subsequent study by Zhu *et al.* described the development of a yeast proteome micro array containing the full-length, purified expression products of over 93 per cent of the organism's complement of 6280 proteincoding genes. A total of 5800 ORFs were cloned as glutathione-S-transferase:: His6 fusions, and expressed in their native cells under a Gal-inducible promoter. Following HT purification, each protein sample was printed in duplicate onto glass slides using a standard robotic microspotter, at a feature density of 13000 samples per array.

This work represented the first systematic cloning and purification of an entire eukaryotic proteome, as well as the first large-scale functional protein array comprising discrete functional proteins. A number of protein attachment chemistries were evaluated, including aldehyde and nickel surface treatment. Aldehyde surfaces promote the covalent attachment of proteins by their N termini, although a concern with this method is that the random crosslinking of primary amines may disrupt the tertiary structures of some proteins.

An alternative approach employed nickel-coated slides, attaching proteins via the incorporated His6 tag. This yielded higher signal-to-noise ratios, presumably because fewer proteins were denatured after nickel attachment and their functional domains were more likely to be oriented uniformly away from the slide surface. Over 90 per cent of the samples were found to yield significant fluorescence signals over background levels, in the range of 10–950 fg of protein.

Several different experiments were performed with the arrays, including a calmodulin-binding survey to assess protein–protein interactions and a large-scale screen for phospholipid-binding specificity. In the latter analysis, self-assembling phosphatidylcholine liposomes were incubated with five different phosphoinositide species and 1 per cent biotin to form a series of biotinylated phospholipid probes. Proteome-wide micro array experiments identified 150 lipidbinding proteins, of which 52 were uncharacterized. In particular, the array-based interaction assays identified sets of proteins that demonstrated preferential binding to one or more phosphoinositides, apart from those that bound all of the phospholipids with equal affinity. As phosphoinositides are important components of membrane structures and secondmessenger pathways, these types of protein–lipid interactions constitute a significant class of biochemical functionality.

FURTHER APPLICATIONS OF PROTEIN ARRAYS

In many cases, complex proteomic experiments may benefit from the combined application of multiple analytical techniques. In a study by Huang *et al.*, a chemical genetic approach was used for the development of a screen for small-molecule inhibitors and enhancers of rapamycin, a peptide exhibiting a variety of functional roles in nutrient metabolism and cell cycle progression in eukaryotes. Following the identification of candidates, yeast proteome arrays were interrogated with two biotinylated rapamycin-inhibitory molecules to identify their potential target proteins. Each was observed to bind multiple target proteins in addition to phosphatidy-linositides.

Analysis of various phenotypes revealed which binding events represented biologically meaningful interactions. A combined experiment involving the use of protein and DNA micro arrays was recently described for the systematic identification and location analysis of transcription factor proteins. Hall *et al.* assayed the yeast proteome in search of novel DNA-binding proteins by probing the protein micro arrays with labeled yeast genomic DNA. A total of 200 DNA-binding proteins were identified, half of which were known, or expected, to bind DNA. Of these, eight candidates were subjected to chromatin profiling via the ChIPchip method, which is designed to identify transcription factor binding sites in genomic DNA. This technique entails the immunoprecipitation of specific protein–DNA complexes using antibodies against a native transcription factor protein or epitope tag. The immunoselected DNA is then sonicated, labeled and used to interrogate a DNA micro array representing intergenic or total genomic regions, thereby revealing the locations of transcription factor binding along a chromosome. ChIP-chip analysis on yeast intergenic arrays revealed Arg5, 6, a mitochondrial enzyme involved in ornithine biosynthesis, to bind DNA at specific nuclear and mitochondrial loci. Altered gene expression levels were observed in Arg5, 6 deletion mutants, further indicating its role as a transcriptional regulator.

NEW CONCEPTS IN CARBOHYDRATE AND PROTEIN NUTRITION

Amino acids are supplied to the duodenum of ruminants by microbial protein synthesized in the rumen, undegraded dietary protein and endogenous protein. Microbial protein usually accounts for a substantial portion of the total amino acids entering the small intestine. Microbial protein is a high quality protein for the animal that is highly digestible in the small intestine. Because microbial growth rates affect amino acid supply to the ruminant animal, it is important to maximize microbial protein synthesis in the rumen. Microbial protein synthesis in the rumen provides the majority of protein supplied to the small intestine of ruminants, accounting for 50 to 80 per cent of total absorbable protein.

The total amount of microbial protein flowing to the small intestine depends on nutrient availability and efficiency of utilization of these nutrients by ruminal bacteria. Ruminal degradation of protein from dietary feed ingredients is one of the most important factors influencing intestinal amino acid supply to ruminants. Proteolysis determines the availability of ammonia nitrogen, amino acids, peptides and branched-chain volatile fatty acids, which influence microbial growth rates in the rumen. Rate and extent of ruminal proteolysis not only affect microbial protein synthesis but also the quantity and quality of undegraded dietary protein that reach the duodenum. Although microbial protein alone may be adequate for low producing ruminants, it can be inadequate for supporting higher levels of growth, wool or milk production. As animal production increases, additional protein must be provided from dietary protein that leaves the rumen undegraded to meet the animal's protein requirement.

The use of high rumen undegradable proteins (RUP) in diets fed to ruminants with high protein requirements can improve the amino acid supply to the animal provided that enough degradable protein is included in the diet to maximize microbial protein. When formulating diets for ruminants, various criteria can be used to select the protein supplement including palatability, ruminal protein degradability, protein quality, intestinal absorption of amino acids, cost per unit of protein, availability and consistency of product, and impact on animal performance. Nutritional models for feeding protein to dairy cattle in the USA have evolved from basic crude protein (NRC, 1978) to more complex systems based on rumen degradable protein (RDP), RUP and intestinal digestion of RUP (NRC, 2001).

The NRC (1989) recognized that intestinal digestion of protein supplements differs; however, empirical data were lacking and as a result, a constant value of 80 per cent was used for all feeds. The main problem was the lack of reliable techniques for estimating intestinal digestion of proteins. With improved techniques, the NRC (2001) has assigned estimates of intestinal digestion to the RUP fraction of each feedstuff. These estimates were obtained from literature using the mobile bag technique and the three-step in situ/*in vitro* procedure of Calsamiglia and Stern (1995). Values used in the French Protein System were adopted for feeds with limited or no data. Feeding proteins to ruminants that are resistant to microbial degradation in the rumen can provide a practical way to increase dietary protein and alter the amino acid profile of the protein reaching the small intestine for digestion and absorption. However, the effects of feeding high RUP on intestinal amino acid supply and animal performance have been inconsistent. Lack of response has been attributed to various factors including depression in ruminal microbial protein synthesis and factors related to quality of the dietary protein such as inadequate or overprotection of protein, reduced intestinal availability of amino acid or inherent amino acid limitations of the dietary protein. A considerable amount of variation among and within feeds in ruminal degradation and intestinal digestion of protein has been reported (Calsamiglia and Stern, 1995; Howie *et al.*, 1996; Yoon *et al.*, 1996). It is important that this variation be considered when determining the value of feeds as sources of protein for the ruminant animal.

Standardized chemical methods of feed analysis were developed over 150 years ago. The proximate analysis system that includes crude protein, crude fibre, ether extract and ash has been in use for over 100 years. However, this system is not adequate in characterizing feed composition relative to rumen and cow needs. Newer chemical and biological methods of feed analysis that better relate to nutritional function have been developed over the past 40 years and continue to be developed. The following describes the rationale and methods used to better characterize feed carbohydrates and protein fractions.

Understanding Carbohydrate Fractions

For the most part, carbohydrates (CHO) are in the diet to support rumen microbial populations. Various populations of microbes are capable of utilizing any CHO compound. In contrast, the cow has intestinal enzymes capable of only digesting sugars and starch. In fact, the cow like other mammals does not really have a CHO requirement as long as other glucose precursors are present in the diet. Therefore, dietary CHO need to be fractionated according to their capability of supporting rumen microbial growth and potential impact on rumen environment.

Carbohydrates are a tremendously diverse group of organic compounds and usually comprise more than 60 per cent of the total diet in most dairy cattle rations. From a nutritional perspective, all CHO provide energy upon oxidation and depending upon their chemical structure may be a precursor for glucose (sugars, starches CHO) or fat (fibre CHO) synthesis. Due to the diversity and complexity in CHO structure, our ability to chemically characterize important nutritional fractions of CHO has been somewhat limited. The challenge for nutritionists has been to be able to adequately quantify different CHO fractions relative to their extent and rate of ruminal degradation or intestinal digestion as to determine their impact on animal performance.

Plant CHOs are primarily differentiated on the basis of their association to the cell wall. Carbohydrates that make up the cell wall are termed structural carbohydrates and are quantified as Neutral Detergent Fibre (NDF). Amount of NDF in a plant is determined by plant species and maturity. Structural CHOs have the property of being slowly fermented, if at all. Therefore, energy yield from these sources would be minimal and slow compared to non-fibre carbohydrates. Due to its slower fermentation rate and need for mastication, NDF is often associated with intake capacity in ruminants. A subset of NDF is acid detergent fibre (ADF), which quantifies the most slowly fermentable or non-fermentable portions of the cell wall. The ADF portion of feeds is often used to estimate digestibility or energy availability.

Non-fibre carbohydrates (NFC) are those compounds not associated with the cell wall, with the exception of neutral detergent soluble fibre (NDSF) compounds (pectic substances, fructosans, beta-glucans). Again, this is a very heterogeneous group of CHO that include organic acids, sugars, starch, and NDSF. A subset of NFC is non-structural CHO (NSC), which are primarily sugars and starches that are very rapidly fermented in the rumen. In contrast to structural CHO, non-fibre CHO can rapidly provide large amounts of energy for microbial protein production or directly to the cow and their presence in a feed increases its digestibility. The diversity of compounds within this nutritionally defined group makes it difficult to directly measure amounts in a feed. Determination of non-fibre CHO is usually by difference, thus accumulating all errors in laboratory analyses for other feed fractions. Within this diverse group of compounds there is tremendous nutritional differences in rate and extent of fermentation or digestion. Ideally, we would like to be able to quantitatively subfractionate this group into organic acids, sugars, starch, and neutral detergent soluble fibre based on differences in nutritional responses. However, procedures to identify some of these non-fibre CHO fractions are only being developed.

Carbohydrate Terminology - The following is a summary list of commonly used CHO terms based on either chemical composition analysis or nutritionally-important fractions.

Crude Fibre (CF) - Original proximate analysis procedure of determining indigestible fibre (cell wall) content of feeds. This procedure incompletely accounts for total cell wall contents, due to limited recovery of hemicellulose and lignin, and thus underestimates total cell wall content of forages and roughages. Crude fibre content will always be less than NDF content and equal to or greater than ADF for a given feed.

Nitrogen Free Extract (NFE) - Proximate analysis procedure to determine readily available carbohydrate sources (sugars and starches). Determined by subtraction: (100 - CP - CF - Ash - EE). Due to the underestimate of fibre by crude fibre, NFE overestimates available carbohydrate, especially for forages.

Neutral Detergent Fibre (NDF) - Van Soest detergent methodology to recover all cell wall components excluding pectin. Contains primarily cellulose, hemicellulose and lignin as well as other resistant non-CHO substances. Correlated with dry matter intake in ruminant animals.

Acid Detergent Fibre (ADF) - Van Soest detergent methodology to recover the indigestible cell wall components. Contains cellulose and lignin as well as heat-damaged protein and other resistant plant compounds. Often associated with feed digestibility.

Non-fibre Carbohydrates (NFC) - Readily digestible (non-cell wall) carbohydrates as determined using the detergent system (organic acids, sugars, starches, soluble fibre). Similar to NFE determination, but always less than NFE for any given feed. Calculated by subtraction: (100 - (NDF + CP + EE + Ash)).

Nonstructural Carbohydrates (NSC) - those carbohydrates that can be digested by mammalian enzymes. Includes organic acids, sugars and starches. Newer chemical fractionation methods are allowing better determinations of starch in feedstuffs.

Lignin - Polyphenol compound, not a carbohydrate, but intimately associated with the plant cell wall carbohydrates. Totally unavailable portion of the cell wall. Lignin increases in amount within a plant structure with increasing plant maturity. Plant species differ in lignin amounts and its impact on carbohydrate availability. Lignin content (per cent DM) times 2.4 is used to estimate the amount of unavailable fibre in a feed.

Effective Fibre or NDF (eNDF, peNDF) - The portion of the total plant cell wall that is effective in increasing rumination and rumen motility. Effectiveness of NDF is based on particle size, degree of lignification, hydration, and density within classes of feeds. Effective fibre will increase rumen pH through its impact on rumination and saliva production.

Factors Affecting Rumen Carbohydrate Availability - A variety of factors beyond chemical composition of CHO can influence rate and extent of ruminal CHO degradation.

1. *Plant Maturity* - As a plant matures, there is an increase in the cell wall content thus diluting out the more digestible components (protein, sugars, minerals; Table). In addition as the plant matures, lignification of cell wall also increases making the cell wall less available for fermentation.
2. *Environmental Conditions* - Rainfall, soil temperature, fertility, cloud cover, location, cutting strategies, etc. all can influence the availability of carbohydrates in the plant. Environmental light, temperature and their interaction have the greatest impact on plant growth. Increased temperature stimulates plant cell wall and lignification reducing plant digestibility. Light exposure will increase soluble CHO content making the plant more digestible.
3. *Processing* - Particle size reduction (grinding) increases surface area

for available microbial attachment and degradation and is very beneficial in increasing cell wall digestion. Steam, extrusion, and popping will alter starch configuration to make it more available. Fermentation (ensiling) will make lesser available carbohydrates more available. Heating can make soluble proteins insoluble and less rumen available.

4. *Plant Species* - Degree of lignification and distribution of lignin within the cell wall will affect rate of digestion of plant carbohydrates. Chemical structure of starch within a feed will dictate rate of fermentation or digestion. Corn starch (amylose) is less digestible than oat or barley starch (amylopectin), but if processed (ensiled, flaked, ground), it can become very available an a potential problem with ruminal acidosis.

Ruminal Nitrogen Metabolism

Ruminal Microbial Protein Synthesis The ultimate goal of proper rumen nutrition is to maximize microbial growth and the amount of RDP that is captured into rumen microbial cells. Maximizing the capture of degradable N not only improves the supply of AA to the small intestine, but also decreases N losses. The rumen is a complex environment inhabited by different microbial species, with each species having different nutrient requirements and metabolism. Therefore, considering the nutrient requirements of ruminal microorganisms is crucial to understanding N metabolism in the rumen as well as the factors that may modify it (Bach *et al.*, 2005).

Factors that affect microbial protein synthesis. Bacteria can utilize carbohydrates and proteins as energy sources. Carbohydrates are the main energy source for bacteria, although they can also be used as carbon skeletons for protein synthesis in combination with ammonia.

Ruminal microbial protein synthesis depends on supply of adequate amounts and type of carbohydrate as an energy source for the synthesis of peptide bonds. Readily fermentable carbohydrates such as starch or sugars are more effective than other sources such as cellulose in promoting microbial growth (Stern and Hoover, 1979). Several *in vitro* (Stern *et al.*, 1978; Henning *et al.*, 1991) and *in vivo* (Casper and Schingoethe, 1989; Cameron *et al.*, 1991) studies demonstrated that infusions of increasing amounts of readily fermentable carbohydrate decreased ammonia-N concentrations due to improved N uptake by ruminal microbes.

However, the optimum ratio of non-fibrous carbohydrates to ammonia-N has not yet been determined. In addition to the importance of the amounts of nutrient supply, the synchrony at which nutrients become available is also important. When rate of protein degradation exceeds the rate of carbohydrate fermentation, large quantities of N can be lost as ammonia, and conversely when the rate of carbohydrate fermentation exceeds protein degradation rate, microbial protein synthesis can decrease (Nocek and Russell, 1988). Whereas the concept of synchronous protein and energy supply has a solid theoretical basis, it is likely that in the complex ecosystem of mixed ruminal microorganisms when nutrient supply is synchronized for a specific subpopulation, it might not be synchronized for other populations. Therefore, average microbial efficiency remains fairly stable. Also, recycled N to the rumen may contribute to stabilize microbial growth even when N supply is not well synchronized (Bach *et al.*, 2005).

Ruminal bacteria are classified as cellulolytic and amylolytic based on their preferential use of energy. Russell *et al.* (1992) proposed a simplified model to describe energy and protein requirements of microbial subpopulations. Microbes that degrade structural carbohydrates (cellulolytic) have low maintenance requirements, grow slowly, and use ammonia-N as their main N source, whereas microorganisms that degrade non-structural carbohydrates (amylolytic) have higher maintenance requirements, grow rapidly, and use ammonia, peptides, and amino acids (AA) as N sources (Russell *et al.*, 1992). However, bacterial growth has been shown to increase with addition of AA and (or) peptides in cellulolytic and amylolytic bacteria (Maeng and Baldwin, 1976; Argyle and Baldwin, 1989; Kernick, 1991). Similarly, fibre digestion was reported to increase with the supply of AA (Griswold *et al.*, 1996; Carro and Miller, 1999) and peptides (Cruz Soto *et al.*, 1994) to pure cellulolytic bacteria.

Atasoglu *et al.* (2001) demonstrated with pure cultures of cellulolytic bacteria, that the incorporation of ammonia-N into microbial cell-N decreased as the proportion of AA increased in the medium, suggesting that cellulolytic bacteria would use AA if available. Similar findings were reported with increasing concentrations of peptides, although Atasoglu *et al.* (2001) reported a greater preference of cellulolytic bacteria for incorporating AA-N compared with peptide-N into their cell-N. However, at typical ruminal peptide and AA concentrations, about 80 per cent of the cell-N is derived from ammonia-N. Addition of branched-chain AA that will ferment to branched-chain VFA, and addition of peptides to ruminal fluid has increased fibre digestion, microbial protein production and growth efficiencies (Russell and Sniffen, 1984; Thomsen, 1985). The increase in microbial growth observed with addition of AA and (or) peptides, may be due to direct incorporation of AA into microbial protein and (or) to increased availability of carbon skeletons from AA deamination, which can be used for energy production or as carbon skeletons for new microbial AA.

Importance of microbial protein to ruminants. The theoretical contribution of microbial protein to the total protein requirement of the lactating dairy cow, calculated at three efficiencies of microbial protein synthesis, is presented in Table. Contribution of microbial protein to total protein requirement was determined using NRC values (2001) for a 680-kg lactating dairy cow producing 25, 35, or 45 kg/d of 4 per cent FCM. At these three milk yields, microbial protein would contribute 51, 49 and 48 per cent, respectively, of the total protein required by the cow when microbial synthesis in the rumen is 30 g of N/kg of organic matter truly digested (OMTD). When milk yield is 45 kg/d, the contribution of microbial protein to total protein required by the cow would increase from 32 to 63 per cent as efficiency of microbial protein synthesis increases from 20 to 40 g of N/kg of OMTD. Stern and Hoover (1979) reviewed the literature and reported that approximately 30 g of N were synthesized per kilogram of OMD in the rumen; values ranged from 10 to 50 g. Efficiencies of microbial protein synthesis used in Table fall into the range of reported values; therefore, calculated contributions of microbial protein clearly depicted the importance of optimizing microbial protein synthesis in the rumen of high producing dairy cows. In addition, these calculations demonstrate that, as milk yield increases, a substantial quantity of RUP from protein supplements must leave the rumen to meet the protein requirement of the cow.

Contribution of microbial protein to growing beef cattle calculated at three efficiencies of microbial protein synthesis is presented in Table. Contribution of microbial protein to total protein requirement was determined using NRC values (1996) for a 250 kg growing beef animal with average daily gains (ADG) of 300, 1200 and 1700 g/day. At these three ADG, microbial protein would contribute 95, 57 and 40 per cent, respectively, of the total protein required by the animal when microbial synthesis in the rumen is 25 g of N/kg of organic matter truly digested (OMTD). Lower efficiencies of microbial protein synthesis, ranging from 20 to 30 g N/kg OMTD, were used because beef cattle are generally less efficient, most probably due to a high population of amylolytic bacteria. Microbial protein contribution to total protein requirement demonstrated a more dramatic difference in beef cattle compared with dairy cattle. This response is probably due to the lack of increase in dry matter intake in beef cattle as ADG increases. However, in both situations, it is clear that substantial quantities of RUP from protein supplements must be incorporated into the diet of high producing ruminants.

Estimates of rumen undegradable protein. Under ideal circumstances, the ruminant nutritionist would prefer to use "book values" for RUP of various protein supplements, such as those found in the NRC (2001) publication. However, ruminal protein degradation is a complex process influenced by various factors such as solubility, protein structure, microbial proteolytic activity, microbial access to the protein and ruminal retention time of dietary protein. There is an inherent variability in reported values for RUP of feeds that is related to conditions under which experiments are conducted; however, part of the variation can also be attributed to differences in quality of protein supplements.

Understanding Protein Fractions

Dietary crude protein (N per cent x 6.25) is not a very useful measure of what potentially happens with consumed nitrogen relative to utilization by the rumen or cow. Crude protein can be chemically separated into fractions on the basis of rumen degradability and solubility and have nutritional relevance to meeting the protein needs of the rumen and cow. Proteins that are rumen degradable would be able to provide nitrogen for microbial protein production. Rumen solubility suggests that the protein source would be more rapidly available. For example, urea, a non-protein nitrogen source, is 100 per cent soluble and degradable and therefore would very rapidly provide ammonia for microbial protein production. Rumen insoluble and very slowly degraded is that dietary protein that can bypass to the abomasum to be digested enzymatically by the cow. The amount of unavailable dietary protein is measured as the amount of nitrogen found in the ADF fraction.

In meeting our goal of maximum microbial protein production, we need to match carbohydrate and protein rates of degradation. This allows for somewhat equivalent amounts of rumen energy and nitrogen availability promoting efficient microbial protein yield and overall dietary protein incorporation. If we can reduce the amount of protein used in the ration and yet maintain or improve milk yield, the dairy cow becomes much more efficient, profitable and environmentally friendly! Protein Terminology - the following is a summary list of commonly used terms relative to protein nutrition of ruminant animals.

- *Crude Protein (CP)*: Total nitrogen per cent of a feed times 6.25 factor. Crude protein analysis does not differentiate between true protein and non-protein nitrogen sources.
- *Digestible Protein (DP)*: Amount of protein absorbed from the intestine from a feed source. Digested protein in the ruminant animal comes from microbial protein and undegraded dietary protein.
- *Soluble Protein (Sol CP, SIP)*: Dietary protein that readily goes into solution in rumen fluid. This is a rapidly available source of NH_3 for the rumen bugs. Composed of NPN and soluble true proteins.
- *Rumen Degradable Protein (RDP, DIP)*: Dietary protein that can be fermented in the rumen by microbes and contribute to the rumen ammonia pool. Rate of passage of digesta through the rumen will determine the extent of which RDP will be degraded.
- *Rumen Undegradable Protein (RUP, UIP), Escape Protein, Bypass Protein:* Protein which is either very slowly degraded in the rumen or which is unavailable to rumen fermentation. This protein is available for digestion in the abomasum to provide amino acids for the cow.
- *Metabolizable Protein (MP)*: Dietary protein (feed and microbial) that has been digested and is available for absorption and utilization.

- *Non-protein Nitrogen (NPN):* Nitrogenous compounds that do not contain linkages of amino acids. Potential protein source for rumen microbes but the utilization is dependent upon dietary CHO availability. Sources include urea, ammonia, biuret and amino acids.
- *Unavailable Protein, Bound Protein (ADIN, ADF-N)*: Protein that is unavailable to rumen degradation or abomasum digestion. This is the amount of nitrogen present in the ADF residue and is a measure of protein bound to fibre due to heat damage. It is very indigestible and reduces feedstuff quality.

Intestinal Protein Digestion

Methods for Measuring Intestinal Protein Digestion The total amount of protein available for absorption from the small intestine depends on the flow of microbial and dietary protein to the duodenum and their respective intestinal digestibilities.

Digestion of protein that leaves the rumen starts in the abomasum with acid-pepsin digestion and is completed in the small intestine with pancreatic and intestinal proteases. The NRC (1989) recognized that intestinal digestion of protein supplements may differ; however, empirical data were lacking, and as a result, a constant value of 80 per cent was used for all feeds. The main problem was the lack of reliable techniques for estimating intestinal digestion of proteins (Stern *et al.*, 1997).

In vivo estimation of intestinal protein digestion involves expensive and labour-intensive experiments and requires the use of surgically prepared animals. Apparent digestion of protein is calculated as the disappearance of CP or amino acids between the duodenum and ileum, which is subject to considerable error associated with digesta sampling, use of digesta flow rate markers, and inherent animal variation. Therefore, several alternative procedures for estimation of intestinal protein digestion in ruminants have been developed.

GROSS PROTEIN REQUIREMENTS

Gross protein requirements have been determined for a few species of fish. Simulated whole egg protein component of test diets contains an excess of indispensable amino acids.

These diets were kept approximately isocaloric by adjusting total protein plus digestible carbohydrate components to a fixed amount as the protein diet treatments were varied over the ranges tested.

Tests in feeding fry, fingerling, and yearling fish have shown that gross protein requirements are highest in initial feeding fry and that they decrease as fish size increases. To grow at the maximum rate, fry must have a diet in which nearly half of the digestible ingredients consist of balanced protein; at 6-8 weeks this requirement is decreased to about 40 per cent of the diet for salmon and trout and to about 35 per cent of the diet for yearling salmonids raised at standard environmental temperature (SET).

Gross protein requirements for young Catfish appear to be less than those for salmonids. Initially feeding fry require that about 50 per cent of the digestible components of the ration be protein, and the requirement decreases with size. Some feeding trials with salmon have indicated direct relationships between changes in the protein requirements of young fish and changes in water temperature.

Chinook salmon in 7°C water require about 40 per cent whole egg protein for maximum growth; the same fish in 15°C water require about 50 per cent protein. Salmon, trout and catfish can use more protein than required for maximum growth because of efficiency in eliminating nitrogenous wastes in the form of soluble ammonia compounds through the gill tissue directly into the water environment.

This system for eliminating nitrogen is more efficient than that available to fowl and mammals. Fowl and mammals consume energy to synthesize urea, uric acid, or other nitrogen compounds which are excreted through the kidney tissue and expelled in urine. Digestible carbohydrate and fat will spare excess protein in the diet as long as the protein requirement for maximum growth is met.

Table. Estimated Dietary Protein Requirement of Certain Fish 1/

Species	Crude Protein Level in Diet for Optimal Growth (g/kg)
Rainbow trout (Salmo gairdneri)	400-460
Carp (Cyprinus carpio)	380
Chinook salmon (Oncorhynchus tshawytscha)	400
Eel (Anguilla japonica)	445
Plaice (Pleuronectes platessa)	500
Gilthead bream (Chrysophrys aurata)	400
Grass carp (Ctenopharyngodon idella)	410-430
Brycon sp.	356
Red sea bream (Chrysophrys major)	550
Yellowtail (Seriola quinqueradiata)	550

Basically the fish must be given a diet containing graded levels of high quality protein and energy and adequate balances of essential fatty acids, vitamins and minerals over a prolonged period. From the resulting dose/response curve the protein requirement is usually obtained by an Almquist plot.

These differences in apparent protein requirement are thought to be due to differences in culture techniques and diet composition. The relatively high dietary protein levels required for maximal growth of certain fish such as grass carp,Ctenopharyngodon idella, and Brycon spp. are surprising as these fish are omnivorous.

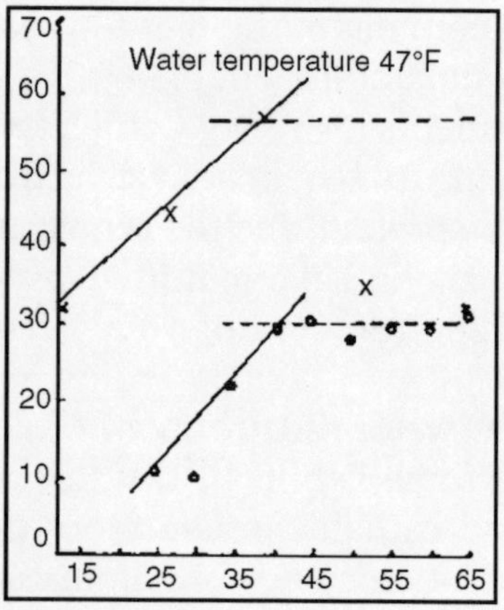

Fig. Protein Requirement of Chinook Salmon at 47°F. Top Curve: Initial Individual Average Weight of Fish, 1.5g. Bottom Curve: Initial Individual average weight of fish, 5.6g.

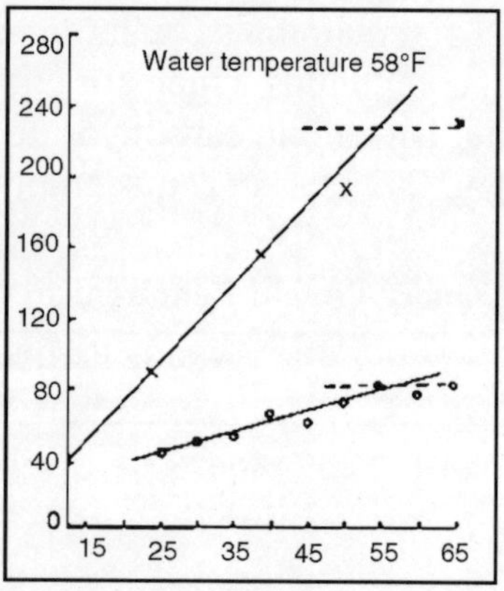

Fig. Protein Requirement of Chinook Salmon at 58°F. Top curve: Initial Individual Average Weight of Fish, 2.6g. Bottom Curve: Initial Individual Average Weight of Fish, 5.8g.

Brycon spp. are grown on unwanted fruit and other plant material of low protein content and under these conditions there is presumably a substantial contribution to their protein intake from a natural food chain. Protein requirement of eurythaline fish such as the rainbow trout, Salmo gairdneri, and the coho salmon, Oncorhynchuskisutch, reared in water of salinity 20 ppt is about the same as the requirement in freshwater. No data are available for the protein requirement of these species in full strength sea water.(35 ppt).

AMINO ACIDS

The amino acids are the building blocks of proteins; about 23 amino acids have been isolated from natural proteins. Ten of these are indispensable for fish. The animal is incapable of synthesizing indispensable amino acids and must therefore obtain these from the diet.

ESSENTIAL AND NON-ESSENTIAL AMINO ACIDS

Salmon, trout and channel catfish fed diets devoid of arginine, histidine, isoleucine, leucine, lysine, methionine, phenylalanine, threonine, tryptophan or valine failed to grow. These same fish fed diets devoid of other L-amino acids grew as well as fish receiving all 18 amino acids tested.

The nitrogen component in the test diets was made up of 18 L-amino acids in the pattern found in whole egg protein. All fish on test recovered rapidly when the missing amino acid was replaced in the diet. The slope of the growth curve of the recovery group was identical with that of fish receiving the complete amino acid test diet. Dispensable amino acids tested were alanine, aspartic acid, cystine, glutamic acid, glycine, proline, serine, and tyrosine. These amino acids were found to be not essential for the growth of salmon, trout and channel catfish.

Quantitative studies on the requirements of the 10 indispensable amino acids used a casein-gelatin mixture supplemented with crystalline L-amino acids. The test diet had an amino acid pattern of 40 per cent whole egg protein for the nitrogen component. Experiments conducted with carp and eel showed a similar lack of growth when an indispensable amino acid was absent from the diet.

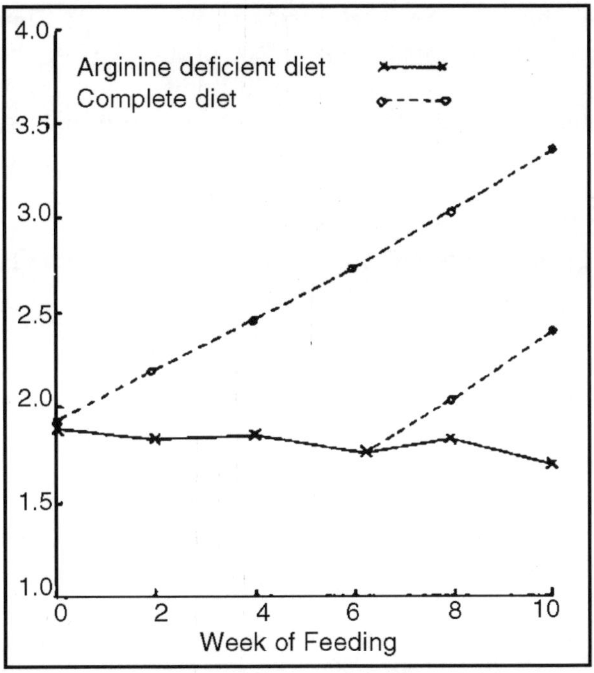

Fig. Growth of Arginine Deficient Fish.

The Deficient group was divided after six weeks on the Deficient diet and the missing amino acid was replaced in one of the two sub-lots.

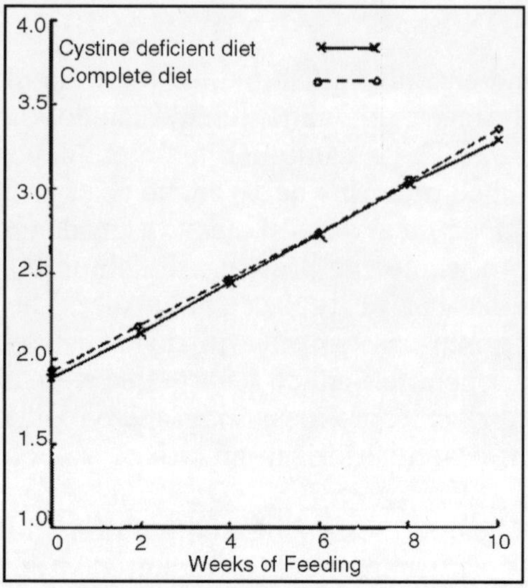

Fig. Growth of Cystine Deficient Fish.

ESSENTIAL AMINO ACIDS AND PROTEIN QUALITY

If the essential amino acid requirements of fish are known, it should be possible to meet these needs in culture systems in a number of ways from different food proteins or combinations of food proteins. Phenylalanine is spared by tyrosine. It is not known to be chemically modified nor rendered unavailable by the harsh conditions to which feedstuff proteins are normally subjected during processing. Measurement of phenylalanine in proteins is uncomplicated so that the provision and evaluation of phenylalanine in proteins in practical diets presents little difficulty.

Lysine is a basic amino acid. In addition to the -amino acid group normally bound in peptide linkage, it also contains a second, -amino group. This -amino group must be free and reactive, otherwise the lysine, although chemically measurable, will not be biologically available. During the processing of feedstuff proteins the -amino group of lysine may react with non-protein molecules present in the feedstuff to form additional compounds that render the lysine biologically unavailable.

Methionine is spared by cystine. However, measurement of the methionine content of feed proteins is not easy as the amino acid is subject to oxidation during processing. After processing, methionine may be present as such or as the sulphoxide or as the sulphone. The sulphoxide may be formed from methionine during acid hydrolysis of the feed protein prior to measurement of its any-no acid composition. Acid hydrolysis of proteins before analysis disturbs the original equilibrium between the two compounds so that the composition of the hydrolysate no longer reflects that of the protein. In determining the methionine content of pure proteins, oxidation of the amino acid to methionine sulphone is normally quantitative. In the case of feed proteins, however, this will not reveal how much methionine or methionine sulphoxide was present in the protein prior to performate oxidation and hydrolysis.

Methionine sulphoxide may have some biological value for fish which may have some capability of reconverting it to methionine and thus partially make up for some of the methionine oxidized during processing.

Methods have recently been reported for measurement of methionine in proteins using an iodoplatinate reagent before and after reduction with titanium trichloride, to give values for both methionine and the sulphoxide in the original protein. A method for measuring methionine specifically by cyanogen bromide cleavage has also been described. Both methods remain to be independently assessed. Microbiological assay of methionine in feed proteins is a valuable tool although there is the danger that oxides of methionine may differ in their activity for micro-organisms and misrepresent values.

QUANTITATIVE REQUIREMENTS OF AMINO ACID

Quantitative requirements by salmonids for the ten indispensable amino acids were determined by feeding linear increments of one amino acid at a time in a test diet containing an amino acid profile identical with whole egg protein except for the amino acid tested. Replicate groups of fish were fed the diet treatments until gross differences appeared in the growth of test lots.

An Almquist plot of growth response indicated the level of amino acids required for maximum growth under those specific test conditions. Diets were designed to contain protein at or slightly below the optimum protein requirement for that species and test condition to assure maximum utilization of the limiting amino acid. A comparison of the requirements for the ten indispensable amino acids between species is shown in Table.

A recent innovation has been the use in test diets of proteins relatively deficient in a given essential amino acid. Thus combinations of fishmeal and zein have been used in test diets to define the requirement of rainbow trout for arginine. Diets containing different relative amounts of casein and gelatin showed that an increase in the level of protein-bound arginine from 11 to 17 g/kg resulted in a significant increase in the growth of channel catfish.

Table. Amino Acid Requirements of Seven Animals 1/

Amino acid	Eel fingerling	Carp fry	Channel catfish	Chinook salmon fingerling	Chick	Young Pig	Rat
Arginine	3.9	4.3		6.0	6.1	1.5	1.0
	(1.7/42)	(1.65/38.5)		(2.4/40)	(1.1/18)	(0.2/13)	(0.2/19)
Histidine	1.9			1.8	1.7	1.5	2.1
	(0.8/42)			(0.7/40)	(0.3/18)	(0.2/13)	(0.4/19)
Isoleucine	3.6	2.6		2.2	4.4	4.6	3.9

	(1.5/42)	(1.0/38.5)		(0.9/41)	(0.8/18)	(0.6/13)	(0.5/13)
Leucine	4.1	3.9		3.9	6.7	4.6	4.5
	(1.7/42)	(1.5/38.5)		(1.6/41)	(1.2/18)	(0.6/13)	(0.9/19)
Lysine	4.8		5.1	5.0	6.1	4.7	5.4
	(2.0/42)		(1.23/24.0)	(2.0/40)	(1.1/18)	(0.65/13)	(1.0/19)
Methionine 2/	4.5	3.1	2.3	4.0	4.4	3.0	3.0
	(2.1/42) 3/	(1.2/38.5)	(0.56/24.0)	(1.6/40)3/	(0.8/18)	(0.6/20)	(0.6/20)
Phenylalanine4/				5.1	7.2	3.6	5.3
				(2.1/41)5/	(1.3/18)	(0.45/13)	(0.9/17)
Threonine	3.6			2.2	3.3	3.0	3.1
	(1.5/42)			(0.9/40)	(0.6/18)	(0.4/13)	(0.2/19)
Tryptophan	1.0			0.5	1.1	0.8	1.0
	(0.4/42)			(0.2/40)	(0.2/18)	(0.2/25)	(0.2/19)
Valine	3.6			3.2	4.4	3.1	3.1
	(1.5/42)			(1.3/40)	(0.8/18)	(0.4/13)	(0.4/13)

Note:

1/ Expressed as per cent of dietary protein. In parentheses, the numerators are requirements as per cent of dry diet, and the denominators are per cent total protein in the diet, 2/ In the absence of cystine, 3/ Methionine plus cystine, 4/ In the absence of tyro sine, 5/ Phenylalanine plus tyrosine.

Arginine requirement of rainbow trout has been determined from a conventional dose/response (growth) curve and also by measuring the tissue (blood and muscle) levels of free arginine in groups of trout given increasing amounts of dietary arginine. After the dietary requirement of the trout for arginine has been met, any further increase in arginine intake led to an increase in the concentration of free arginine in blood and muscle. Good agreement was obtained between the two methods. The data shown in Table suggest that real differences exist between fish species in their requirement for certain amino acids. This leads to difficulties in formulating the protein component of practical diets for those species whose amino acid requirements are not yet known. A possible solution is to use, for each amino acid, the highest level required by any of those species for which data is available. The need for further quantitative data on the amino acid requirements of fish, especially those actually or potentially useful as farm animals, is obvious.

SUPPLEMENTING DIETS WITH AMINO ACIDS

One solution to the use of proteins that are relatively deficient in one or more amino acids is to supplement the protein with appropriate amounts of the amino acid needed in practical diets. Fish appear to utilize free amino acids at various degrees of efficiency.

Young carp, Cyprinus carpio, were shown to be unable to grow on diets in which the protein component (casein, gelatin) was replaced by a mixture of amino acids similar in overall composition. A trypsin hydrolyzate of casein was equally ineffective. However, if a diet containing free amino acids as the protein component is carefully neutralized with NaOH to pH 6.5-6.7 then some growth of young carp does occur. This growth was markedly inferior to that occurring on a comparable casein diet under the same conditions. Channel catfish are also unable to utilize free amino acids given as supplements to deficient proteins. When soybean meal was substituted isonitrogenously for menhaden meal, growth and feed efficiency of channel catfish were substantially reduced. Addition of free methionine, cystine or lysine, the most limiting amino acids, to these soy-substituted diets did not enhance weight gain. Raising the arginine level of catfish diets from 11 to 17 g/kg by isonitrogenous substitution of gelatin for casein enhanced weight gain significantly but the addition of free arginine, cystine, tryptophan or methionine to casein had little effect on growth or food conversion.

Salmonids are able to utilize free amino acids for growth. A zein-gelatin diet supplemented with lysine and trytophan was shown to be markedly superior to an unsupplemented zein-gelatin diet for rainbow trout when weight gain and protein utilization were used as criteria. Several investigators have demonstrated the potential of supplementing amino acid deficient proteins with limiting amino acids in diets for salmonids. Casein supplemented with six amino acids produced feed conversion ratios with Atlantic salmon similar to those obtained when an isolated fish protein was used as the dietary protein source.

Soybean meal supplemented with five or more amino acids (including methionine and lysine) was a superior protein source to soybean meal alone for rainbow trout. Single additions of methionine and lysine did not, however, improve the value of soybean meal. These results suggest that the amino acid spectrum of the isolated fish protein they used may possibly approximate the amino acid requirement of rainbow trout. The nutritional value of a soy protein isolate could be enhanced by supplementing it with the first limiting amino acid; *i.e.*, methionine.

Diets containing, as protein component, fishmeal, meat and bone meal, and yeast and soybean meal could be improved by supplementing with cystine (10 g/kg) and tryptophan (5 g/kg) together. Fishmeal can be entirely replaced without a reduction in food conversion rate in diets for rainbow trout by a mixture of poultry by-product meal and feather meal together with 17 g lysine HCL/kg, 4.8 g DL-methionine/kg, and 1.44 g DL-tryptophan/kg.

THE ROLE OF CARBOHYDRATES IN NUTRITION

Carbohydrates are polyhydroxy aldehydes, ketones, alcohols, acids, their simple derivatives and their polymers having linkages of the acetal type. They may be classified according to their degree of polymerization and may be divided initially into three principal groups, namely sugars, oligosaccharides and polysaccharides.

Each of these three groups may be subdivided on the basis of the monosaccharide composition of the individual carbohydrates. Sugars comprise monosaccharides, disacc-harides and polyols (sugar alcohols); oligosaccharides include malto-oligosaccharides, principally those occurring from the hydrolysis of starch, and other oligosaccharides, *e.g.*, a -galactosides (raffinose, stachyose, etc.) and fructo-oligosaccharides; the final group are the polysaccharides which may be divided into starch (a -glucans) and non-starch polysaccharides of which the major components are the polysaccharides of the plant cell wall such as cellulose, hemicellulose and pectin (2,3,4).

Table. The Major Dietary Carbohdrates

Class (DP*)	Sub-Group	Components
Sugars (1-2)	Monosaccharides	Glucose, galactose, fructose
	Disaccharides	Sucrose, lactose, trehalose
	Polyols	Sorbitol, mannitol
Oligosaccharides (3-9)	Malto-oligosaccharides	Maltodextrins
	Other oligosaccharides	Raffinose, stachyose, fructo-oligosaccharides
Polisaccharides (>9)	Starch	Amylose, amylopectin, modified starches
	Non-starch polisaccharides	Cellulose, hemicellulose, pectins, hydrocolloids

DP * = Degree of polymerization

TOTAL CARBOHYDRATE

Although the individual components of dietary carbohydrate are readily identifiable, there is some confusion as to what comprises total carbohydrate as reported in food tables. Two principal measures of total carbohydrate are used, firstly, that derived by "difference" and secondly the direct measurement of the individual components which are then combined to give a total. Calculating carbohydrates by "difference" has been used since the turn of the century.

The protein, fat, ash and moisture content of a food are determined, subtracted from the total weight of the food and the remainder, or "difference", is considered to be carbohydrate. There are, however, a number of problems with this approach to total carbohydrate analysis in that the "by difference" figure includes a number of non-carbohydrate components such as lignin, organic acids, tannins, waxes, and some Maillard products. In addition to this error, it combines all of the analytical errors from the other analyses. Finally, a single global figure for carbohydrates in food is uninformative because it fails to identify the many types of carbohydrates in a food and thus to allow some understanding of the potential physiological properties of those carbohydrates.

TERMINLOGY

In deciding how to classify dietary carbohydrate the principal problem is to reconcile the various chemical divisions of carbohydrate with that which reflects physiology and health. A classification based purely on chemistry does not allow a ready translation into nutritional terms since each of the major classes of carbohydrate have a variety of physiological effects.

However, a classification based on physiological properties also creates a number of problems in that it requires a single effect to be considered as overridingly important and to be used as the basis of the classification. This dichotomy has led to the introduction of a number of terms to describe various fractions and sub-fractions of carbohydrate.

Sugars

The term "sugars" is conventionally used to describe the mono and disaccharides. "Sugar", by contrast, is used to describe purified sucrose as are the terms "refined sugar" and "added sugar".

Extrinsic and Intrinsic Sugars

These terms had their origin in a United Kingdom Department of Health committee in 1989, which was looking at the question of sugars in the diet. The terms were developed to help the consumer choose between what were considered to be healthy sugars and those which were not. Intrinsic sugars were defined as sugars occurring within the cell walls of plants, *i.e.*, naturally occurring, while extrinsic sugars were those which were usually added to foods.

Because lactose in milk is also an extrinsic sugar, an additional phrase "non-milk extrinsic sugars" was developed. These terms have not gained wide acceptance either in the U.K. or other countries in the world. There are no current plans to measure these sugars separately in the diet nor to incorporate their use into food tables.

Complex Carbohydrates

This term was first used in the McGovern report, "Dietary Goals for the United States" in 1977. The term was coined largely to distinguish sugars from other carbohydrates and in the report denotes "fruit, vegetables and whole-grains". The term has since come to be used to describe either starch alone, or the combination of all polysaccharides. It was used to encourage consumption of what were considered to be healthy foods such as whole-grain cereals, etc., but becomes meaningless when used to describe fruit and vegetables which are low in starch. Furthermore, it is now realised that starch, which is by any definition a complex carbohydrate, is variable metabolically with some forms being rapidly absorbed and having a high glycemic index and some being resistant to digestion. The term "complex carbohydrate" has encompassed, at various times, starch, dietary fibre and non-digestible oligosaccharides. As a substitute term for starch, however, it would seem to have little merit and, in principle, it is better to discuss carbohydrate components by using their common chemical names.

Available and Unavailable Carbohydrate

A major step forward conceptually in our understanding of carbohydrates was made by McCance and Lawrence in 1929 with the division of dietary carbohydrate into available and unavailable. In an attempt to prepare food tables for diabetic diets they realised that not all carbohydrates could be "utilized and metabolized", *i.e.*, provide the body with "carbohydrates for metabolism". Available carbohydrate was defined as "starch and soluble sugars" and unavailable as "mainly hemicellulose and fibre". This concept proved useful, not the least because it drew attention to the fact that some carbohydrate is not digested and absorbed in the small intestine but rather reaches the large bowel where it is fermented. It suggests that the site of digestion or fermentation in the gut of carbohydrate is of overriding importance. However, it is misleading to talk of carbohydrate as "unavailable" because some indigestible carbohydrate is able to provide the body with energy through fermentation.

There are many properties of carbohydrate of which digestibility and fermentability are only two. A more appropriate substitute for the terms "available" and "unavailable" today would be to describe carbohydrates as either as glycemic or non-glycemic, which is closer to the original concept of McCance and Lawrence.

Resistant Starch

One of the major developments in our understanding of the importance of carbohydrates for health in the past twenty years has been the discovery of resistant starch. Resistant starch is defined as "starch and starch degradation products not absorbed in the small intestine of healthy humans". The main forms of resistant starch are physically enclosed starch, *e.g.*, within intact cell structures (RS_1), some raw starch granules (RS_2) and retrograded amylose (RS_3).

Modified Starch

The proportions of amylose and amylopectin in a starchy food is variable and can be altered by plant breeding. Techniques using genetic engineering are rapidly emerging, enabling starches to be produced for specific purposes by genetically modifying the crop used for their production. High amylose corn starch and high amylopectin (waxy) corn starch have been available for a long time, and display quite different functional as well as nutritional properties. High amylose starches require higher temperatures for gelatinization and are more prone to retrograde and to form amylose-lipid complexes. Such properties can be utilized in the formulation of foods with low glycemic index and/or high resistant starch content. Physical modifications of starches include pregelatinization and partial hydrolysis (dextrinization).

Chemical modification is mainly the introduction of side groups and cross-linking or oxidation. These modifications may be used to decrease viscosity and to improve gel stability, mouthfeel, appearance and texture, and resistance for heat treatment. The application of modified starches as fat replacers is another important area. Some modified starches may be partly resistant to digestion in the small intestine, thereby adding to resistant starch.

Dietary Fibre

The original description of dietary fibre by Trowell in 1972 was "that portion of food which is derived from cellular walls of plants which is digested very poorly by human beings". This is not an exact description of any carbohydrate in the diet but is more a physiological concept. It was linked by Burkitt and Trowell to the etiology of a number of "Western diseases" and on the basis of this a hypothesis relating fibre to health was developed. The use of the term has, however, caused many difficulties over the years because of controversies regarding definition. Moreover, the proposal that there are a number of dietary fibre deficiency disorders is an over-simplification and needs to be modified now in the light of new knowledge of diet and disease. The main components of dietary fibre are derived from the cell walls of plant material in the diet and comprise cellulose, hemicellulose and pectin. Lignin, a non-carbohydrate component of the cell wall is also often included. Dietary fibre is a term which is felt to be valuable for the consumer who looks upon this as a healthy component of the diet. At the present time there is no consensus as to which components of carbohydrate should be included as dietary fibre and different authors have variously included non-starch polysaccharides and resistant starch. More recently it has been suggested that non-digestible oligosaccharides should also be included. Dietary fibre has also been defined by method. While there is general agreement that the non-starch polysaccharides are the principal part of dietary fibre there is currently no consensus as to whether other components should be included in this term. It has been suggested that the use of the term dietary fibre be gradually phased out. Its widespread use and popularity with the consumer has made this difficult in practice and the term has been useful in nutrition education and product development.

Soluble and Insoluble Fibre

These terms developed out of the early chemistry of non-starch polysaccharides which showed that the fractional extraction of these polysaccharides could be controlled by changing the pH of solutions. They proved very useful in the initial understanding of the physiological properties of dietary fibre, allowing a simple division into those which principally had effects on glucose and lipid absorption from the small intestine and those which were slowly and incompletely fermented and had more pronounced effects on bowel habit. However, the separation of soluble and insoluble fractions is not chemically very distinct being dependent on the conditions of extraction. Moreover, the physiological differences are not, in fact, so distinct with much insoluble fibre being rapidly and completely fermented while not all soluble fibre has effects on glucose and lipid absorption.

NUTRITION AND GROWTH OF BACTERIA

Every organism must find in its environment all of the substances required for energy generation and cellular biosynthesis. The chemicals and elements of this environment that are utilized for bacterial growth are referred to as nutrientsor nutritional requirements. Many bacteria can be grown the laboratory inculture media which are designed to provide all the essential nutrients in solution for bacterial growth. Bacteria that are symbionts or obligate intracellular parasites of other cells, usually eucaryotic cells, are (not unexpectedly) difficult to grow outside of their natural host cells. Whether the microbe is a mutualist or parasite, the host cell must ultimately provide the nutritional requirements of its resident. Many bacteria can be identified in the environment by inspection or using genetic techniques, but attempts to isolate and grow them in artificial culture has been unsuccessful. This, in part, is the basis of the estimate that we may know less than one per cent of all procaryotes that exist.

TRACE ELEMENTS

Table ignores the occurrence of trace elements in bacterial nutrition. Trace elements are metal ions required by certain cells in such small amounts that it is difficult to detect (measure) them, and it is not necessary to add them to culture media as nutrients. Trace elements are required in such small amounts that they are present as "contaminants" of the water or other media components. As metal ions, the trace elements usually act as cofactors for essential enzymatic reactions in the cell. One organism's trace element may be another's required element andp vice-versa, but the usual cations that qualify as trace elements in bacterial nutrition are Mn, Co, Zn, Cu, and Mo.

CARBON AND ENERGY SOURCES FOR BACTERIAL GROWTH

In order to grow in nature or in the laboratory, a bacterium must have an energy source, a source of carbon and other required nutrients, and a permissive range of physical conditions such as O_2 concentration, temperature, and pH. Sometimes bacteria are referred to as individuals or groups based on their patterns of growth under various chemical (nutritional) or physical conditions. For example, phototrophs are organisms that use light as an energy source; anaerobes are organisms that grow without oxygen; thermophiles are organisms that grow at high temperatures. All living organisms require a source of energy. Organisms that use radiant energy (light) are called phototrophs. Organisms that use (oxidize) an organic form of carbon are called heterotrophs or (chemo)heterotrophs. Organisms that oxidize inorganic compounds are called lithotrophs. The carbon requirements of organisms must be met by organic carbon (a chemical compound with a carbon-hydrogen bond) or by CO_2. Organisms that use organic carbon are heterotrophs and organisms that use CO_2 as a sole source of carbon for growth are called autotrophs. Thus, on the basis of carbon and energy sources for growth four major nutritional types of procaryotes may be defined.

Table. Major Nutritional Types of Procaryotes

Nutritional Type	Energy Source	Carbon Source	Examples
Photoautotroph	Light	CO_2 Green Bacteria	Cyanobacteria, some Purple and
Photoheterotrophs	Light	Organic compounds	Some Purple and Green Bacteria
Chemoautotrophs or Lithotrophs Lithoautotrophs)	Inorganic *e.g.* compounds, H_2, NH_3, NO_2, H_2S	CO_2	A few Bacteria and many Archaea
Chemoheterotrophs Heterotrophs	Organic compounds	Organic compounds	Most Bacteria, some or Archaea

Almost all eucaryotes are either photoautotrophic (*e.g.* plants and algae) or heterotrophic (*e.g.* animals, protozoa, fungi). Lithotrophy is unique to procaryotes and photoheterotrophy, common in the Purple and Green Bacteria, occurs only in a very few eucaryotic algae.

Phototrophy has not been found in the Archaea, except for non-photosynthetic light-driven ATP synthesis in the extreme halophiles.

GROWTH FACTORS

This simplified scheme for use of carbon, either organic carbon or CO_2, ignores the possibility that an organism, whether it is an autotroph or a heterotroph, may require small amounts of certain organic compounds for growth because they are essential substances that the organism is unable to synthesize from available nutrients. Such compounds are called growth factors.

Growth factors are required in small amounts by cells because they fulfill specific roles in biosynthesis.

The need for a growth factor results from either a blocked or missing metabolic pathway in the cells.

Growth factors are organized into three categories:

1. Purines and pyrimidines: required for synthesis of nucleic acids (DNA and RNA)
2. *Amino Acids:* Required for the synthesis of proteins
3. *Vitamins:* Needed as coenzymes and functional groups of certain enzymes

Some bacteria (*e.g.* E. coli) do not require any growth factors: they can synthesize all essential purines, pyrimidines, amino acids and vitamins, starting with their carbon source, as part of their own intermediary metabolism.

Certain other bacteria (*e.g.* Lactobacillus) require purines, pyrimidines, vitamins and several amino acids in order to grow. These compounds must be added in advance to culture media that are used to grow these bacteria.

The growth factors are not metabolized directly as sources of carbon or energy, rather they are assimilated by cells to fulfill their specific role in metabolism. Mutant strains of bacteria that require some growth factor not needed by the wild type (parent) strain are referred to as auxotrophs.

Thus, a strain of E. coli that requires the amino acid tryptophan in order to grow would be called a tryptophan auxotroph and would be designated E. colitroph.

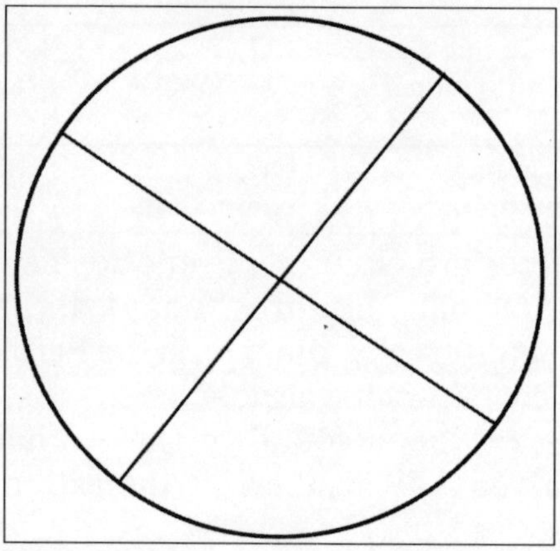

CULTURE MEDIA FOR THE GROWTH OF BACTERIA

For any bacterium to be propagated for any purpose it is necessary to provide the appropriate biochemical and biophysical environment. The biochemical (nutritional) environment is made available as a culture medium, and depending upon the special needs of particular bacteria (as well as particular investigators) a large variety and types of culture media have been developed with different purposes and uses. Culture media are employed in the isolation and maintenance of pure cultures of bacteria and are also used for identification of bacteria according to their biochemical and physiological properties.

The manner in which bacteria are cultivated, and the purpose of culture media, varies widely. Liquid media are used for growth of pure batch cultures, while solidified media are used widely for the isolation of pure cultures, for estimating viable bacterial populations, and a variety of other purposes.

The usual gelling agent for solid or semisolid medium is agar, a hydrocolloid derived from red algae. Agar is used because of its unique physical properties (it melts at 100°C and remains liquid until cooled to 40°C, the temperature at which it gels) and because it cannot be metabolized by most bacteria. Hence as a medium component it is relatively inert; it simply holds (gels) nutrients that are in aqueous solution.

TYPES OF CULTURE MEDIA

Culture media may be classified into several categories depending on their composition or use. A chemically-defined (synthetic) medium is one in which the exact chemical composition is known. A complex (undefined) medium is one in which the exact chemical constitution of the medium is not known. Defined media are usually composed of pure biochemicals off the shelf; complex media usually contain complex materials of biological origin such as blood or milk or yeast extract or beef extract, the exact chemical composition of which is obviously undetermined. A defined medium is a minimal medium if it provides only the exact nutrients (including any growth factors) needed by the organism for growth.

The use of defined minimal media requires the investigator to know the exact nutritional requirements of the organisms in question. Chemically-defined media are of value in studying the minimal nutritional requirements of microorganisms, for enrichment cultures, and for a wide variety of physiological studies. Complex media usually provide the full range of growth factors that may be required by an organism so they may be more handily used to cultivate unknown bacteria or bacteria whose nutritional requirement are complex (*i.e.*, organisms that require a lot of growth factors, known or unknown). Complex media are usually used for cultivation of bacterial pathogens and other fastidious bacteria.

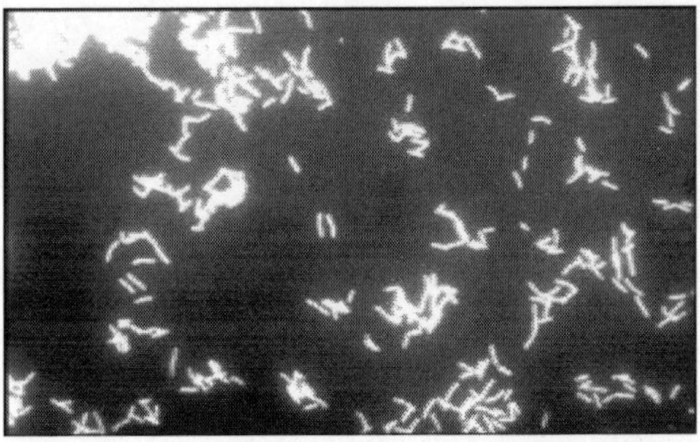

Most pathogenic bacteria of animals, which have adapted themselves to growth in animal tissues, require complex media for their growth. Blood, serum and tissue extracts are frequently added to culture media for the cultivation of pathogens. Even so, for a few fastidious pathogens such as Treponema pallidum, the agent of syphilis, and Mycobacterium leprae, the cause of leprosy, artificial culture media and conditions have not been established. This fact thwarts the the ability to do basic research on these pathogens and the diseases that they cause. Other concepts employed in the construction of culture media are the principles of selection and enrichment. A selective medium is one which has a component(s) added to it which will inhibit or prevent the growth of certain types or species of bacteria and/or promote the growth of desired species. One can also adjust the physical conditions of a culture medium, such as pH and temperature, to render it selective for organisms that are able to grow under these certain conditions.

A culture medium may also be a differential medium if allows the investigator to distinguish between different types of bacteria based on some observable trait in their pattern of growth on the medium. Thus a selective, differential medium for the isolation of Staphylococcus aureus, the most common bacterial pathogen of humans, contains a very high concentration of salt (which the staph will tolerate) that inhibits most other bacteria, mannitol as a source of fermentable sugar, and a pH indicator dye. From clinical specimens, only staph will grow. S. aureus is differentiated from S. epidermidis (a non-pathogenic component of the normal flora) on the basis of its ability to ferment mannitol. Mannitol-fermenting colonies (S. aureus) produce acid which reacts with the indicator dye forming a coloured halo around the colonies; mannitol non-fermenters (S. epidermidis) use other non-fermentative substrates in the medium for growth and do not form a halo around their colonies.

An enrichment medium employs a slightly different twist. An enrichment medium contains some component that permits the growth of specific types or species of bacteria, usually because they alone can utilize the component from their environment. However, an enrichment medium may have selective features. An enrichment medium for nonsymbiotic nitrogen-fixing bacteria omits a source of added nitrogen to the medium. The medium is inoculated with a potential source of these bacteria (*e.g.* a soil sample) and incubated in the atmosphere wherein the only source of nitrogen available is N_2. A selective enrichment medium for growth of the extreme halophile (Halococcus) contains nearly 25 per cent salt [NaCl], which is required by the extreme halophile and which inhibits the growth of all other procaryotes.

6

Domestic Animal Diversity

Domestic animal diversity (DAD) is rightly labeled one of the most threatened aspects of biodiversity by the Food and Agriculture Organization (FAO), the UN entity charged with global oversight of DAD documentation and conservation. Much of today's remaining diversity in domestic animal breeds survives in traditional farming and herding communities in the South, where it was generated by local/indigenous knowledge and social organization. Yet FAO and other international organizations have made little effort to integrate such knowledge and practice into their global strategies for understanding and maintaining DAD.

Their rationale for saving local/indigenous breeds from impending extinction seems to lie mainly in these animals' possession of valuable genetic material that may be of potential benefit to the North or to humanity at large. (Ironically, many indigenous breeds are now at risk due to cross-breeding policies previously promoted by the same formal-sector institutions now seeking to save them.) Scant attention has been paid to the fact that endangered breeds are frequently associated with marginalized social groups whose economic and cultural survival depends directly upon these animals - and who thus have an even stronger and more immediate interest in their conservation.

Agro-biodiversity is composed of crop and livestock diversity at the levels of genes, species and habitats. In general parlance, however, farm-animal diversity refers to the recognized breeds of ruminants, swine, equines and poultry that humans have developed across the millennia from less than a score of wild animal species. This DAD is now at grave risk. According to the Food and Agriculture Organization, only some 5,000 breeds or strains of all species of farm-animal domesticates still exist, from among untold thousands developed historically. Europe alone has lost nearly half the breeds of farm animals found there at the beginning of the twentieth century. Worldwide, about a third of the remaining 5,000 are in danger of extinction, with breeds disappearing at the rate of more than one per week.

In the North, most livestock is raised under intensive or even industrialized production systems that make for impressive outputs of meat, milk, eggs and fibre. These outputs have been achieved by using such high-tech reproductive operations as artificial insemination and embryo transplantation, selecting solely for productivity without regard to other traits such as fitness. In the process, the genetic base of the species so treated has been greatly narrowed and their natural survival traits have been compromised. The resulting food-producing 'machines' are very vulnerable to disease and generally less hardy. They therefore require high veterinary and other inputs, such as special feeds, expensive housing and sophisticated husbandry. Further, these industrial breeds' fertility and reproductive performance have been undermined to such an extent that they often have difficulty mating, giving birth and mothering their young. It is questionable whether they could even survive outside the managed environments and factory farms in which they are raised.

To ensure at least a modicum of fitness and vitality in future populations of food-producing animals, and to keep genetic options open, access to fresh genetic material will always be required. But most of the wild relatives of today's animal domesticates are extinct. This means that a major source of such material lies in the more rustic livestock kept by farmers and herders under extensive, subsistence-oriented production systems in the South.

Variously termed local/indigenous or unimproved (that is, by Northerners), these Southern landraces of animals still harbour such invaluable characteristics as natural disease-resistance, strength, hardiness, good libido and fertility, and other survival-oriented traits. Indeed, in the absence of livestock with these traits, the majority of the world's people could not survive. Most of the world population is still rural, and all rural groups rely on domesticated animals for a key part of their livelihood; but few can afford or access the high-tech inputs and breeds typical in the North. Fortunately for people everywhere, however, Southerners' landraces often embody special traits that are of present or future economic importance. Their genetic makeup may offer possibilities for tastier, healthier or more specialized foods than the bland, generic products from factory animals.

But if, like their wild relatives, indigenous livestock breeds also disappear, then future human generations will have no genetic manoeuvring room in which to adapt their food-animal production systems to emerging disease threats or changing environmental and economic conditions. Thus serious DAD conservation efforts by FAO and other formal-sector institutions like the International Livestock Research Institute (ILRI) are vital to future food security.

Article 8 of the UN Convention on Biological Diversity states that genetic resources should be conserved in the 'surroundings where they have developed their distinct properties'. For livestock, this means the pastoral and farming communities that nurture most of the world's existing DAD. The diversification of a livestock species into many different strains or breeds is the outcome of different ethnic and social groups' managing that species in a wide variety of habitats and manipulating its genetic composition according to localized biophysical (climate, disease and predator threats, availability of forage, water, minerals, etc.) and socio-cultural (*e.g.* knowledge, beliefs, labour, social organization) conditions in the context of these human groups' varying needs and preferences for livestock goods and services.

In other words, albeit without the North's high tech, people in the South and in traditional societies everywhere have also consciously shaped breeds to their own ends - contrary to ethnocentric beliefs among many animal scientists that livestock in developing countries evolved without human intervention. Although targeted studies of animal breeding in traditional cultures are rare, recent overviews of local/indigenous knowledge of and practices in animal reproduction, breeding and other husbandry arenas attest to a wide range of astute interventions in livestock genetics. Given their direct dependence on livestock for daily survival, nomadic pastoralists in particular have accumulated a rich store of knowledge and expertise in animal breeding.

KNOWLEDGE OF BREEDING

All long-time stock raising peoples have a practical, working knowledge of genetics, at least at a phenotypic level. Many pastoral and agro-pastoral peoples also keep detailed mental or oral stock records. Indeed, even children can often identify the pedigrees of all the animals in their own and nearby families' herds. This is because an animal's ancestry is often encoded in its name, and names are not changed when animals are sold or exchanged. Usually, records follow female bloodlines.

Furthermore, many pastoral peoples restrict or even taboo the sale of (especially female) breeding stock outside their home community or ethnic segment (tribe, caste). They may do so because they consider such animals their prime capital, as do practically all camel pastoralists or ideological explanations may be offered. For example, Andean agro-pastoralists' believe that the souls of animals slaughtered far from their home community will be unable to make their way back to be re-born, thus prejudicing continued herd reproduction. Whatever the emic reason, etically such social restrictions isolate animal populations, and thus encourage and stabilize the development of unique, localized breeds.

Breeding Goals

Nearly every traditionally stock raising society has developed one or more distinctive livestock breeds to suit its particular environment and animal-product needs and wants. People often have multiple breeding goals for a given species. Still, the first consideration is sheer survivability. Thus stock are selected and bred for: resistance to dangerous endemic diseases; general hardiness, as in an ability to survive even when forage, water and minerals are scarce and to trek long distances in search of them; and adaptability to local climatic extremes, whether in temperature, precipitation or altitude. Also often figuring in breeding choices is animal behaviour and temperament as these relate to livestock survival and human management thereof, for example, good mothering instincts and aggressiveness towards predators, but also herd ability, tameness, and loyalty, the latter reflected in stock finding their own way home and resisting contact with humans other than their masters/mistresses.

Of course, in the absence of high-tech operations, good libido and natural fertility are general desiderata. Other breeding goals may involve the particular goods and services sought from a given species or animal, beyond just maximum product yields. To take a few examples: strength or fleetness in draught, pack or riding animals; intelligence in lead or work animals; and particular fibre Colour or qualities, milk-fat content or eggshell hardness, according to market demands and conditions. Finally, various aesthetic and cultural considerations may also be at work, as when a people prize certain phenotypic characteristics (*e.g.* a particular horn or body conformation, feather configuration or physiognomic feature) for special social or ceremonial purposes.

Breeding Practices

These consist of the techniques and social institutions by which breeding goals are achieved. The most common is simple selection of animals allowed to mate. As a rule of thumb, traditional stock raisers focus selection on male animals; although typically they also take into account the male's female forebears and relatives.

Selection is negatively implemented by such techniques as: castration - probably the most common way of preventing unsuitable male animals from breeding; culling, slaughtering or selling animals considered unfit as breeding stock; subdividing herds by sex and age; tying aprons or sheaths over the genitalia; and in one culture, inserting stones into the vagina as a sort of IUD. It is positively implemented by, for example, corralling together or manually coupling prime mating pairs; purchasing, hiring or borrowing preferred studs; 'stealth' techniques such as grazing or losing ones' female animals in another's herd that has desirable studs; and other social mechanisms such as multi-household stud ownership.

The practice of offspring testing is reported for camel breeders in Somalia and India. Also widely used in scientific breeding, offspring testing refers to restricting a male animal's mating until his first crop of progeny can be evaluated. Only if his first offspring live up to expectations will the male then be used more widely.

Some stock raising societies fastidiously avoid inbreeding, while others see no harm in it. Attitudes about animal inbreeding sometimes mirror local marriage practices. For instance Arabian Bedouin are endogamous and so permit inbreeding, whereas the exogamous pastoralists of Rajasthan never allow it. Indeed, breeding practices are often intertwined with cultural beliefs and local social organization. To take another example, the historically superior quality of cattle in the Marwar region of Rajasthan has been attributed to the fact that all sub-standard bulls were systematically castrated. This practice hinged on the availability of low-caste people to perform this socially debasing task. Nowadays, however, few such people are willing to do this work, and cattle quality has thus suffered.

Breeds

Many local/indigenous livestock species and breeds have a close association with particular ethnic groups with whom they have co-evolved. Indeed, these groups may couch their very identity in terms of the human/animal bond. The survival of the two - animals and humans - is interlinked. Without special interventions, animals will seldom retain their genetic uniqueness outside the traditional production systems that gave them rise. Conversely, the ethnic group in question cannot continue their production system and thus their cultural (or even physical) survival without these animals. A few examples follow, drawn just from among stock raising peoples of a single country, India.

- One of the best indigenous cattle breeds is the Sahiwal, which combines exceptional hardiness with high milk yields. Muslim pastoralists are the hereditary keepers of Sahiwal, which they carefully maintain as purebreds by vigourously resisting all attempts by government and development agencies to introduce cross-breeding.
- Rebari pastoralists developed the Gir and Kankrej cattle breeds, which became the founding stock of famous beef cattle breeds in Brazil and the southern US But their original creators are being pushed off their pasturelands, thereby imperilling these breeds' continued existence in their original surroundings.
- A hereditary camel-breeding caste, the Raika can no longer earn a living from their breeding work because alienation of pastures has reduced the sizes and reproductive rates of their herds. But regarding themselves as guardians of the camel, some of them continue to keep camels anyway.
- Bastar tribals in Madhya Pradesh are the only group to have protected the rare Aseel chicken, which is extinct everywhere else.

Such examples could be multiplied many times over across the globe. The point is that the human groups who have nurtured and developed special animal genetic resources should be regarded as their rightful owners. As such, these peoples can play, and should be accorded, a vital key role in initiatives to conserve these invaluable resources.

Conserving Livestock Diversity

FAO has a global mandate to study, advice and set guidelines on conserving livestock genetic resources for present and future food security. FAO's Initiative for Domestic Animal Diversity began by establishing a database known as the DAD Information System. The system is designed to inventory and monitor DAD worldwide. The Initiative further seeks to conserve those breeds classified in the database as endangered and critical, and to promote exchange of these precious genetic resources. To these ends, the Initiative has established: an intergovernmental mechanism, a technical programme of management support for UN member nations, a cadre of experts, and a country-based global infrastructure of national coordinators.

On paper, FAO recognizes the importance of stock raisers' local/indigenous breeding knowledge. 'The indigenous knowledge that has helped to produce and maintain domestic animal diversity is largely unexplored and yet this knowledge is essential in order to understand and continue developing these animal genetic resources'. This recognition has not been translated into any specific activities, however.

Data Basing

So far, no efforts have been made to recover local/indigenous knowledge about breeds and then integrate it into the DAD-IS. Database information is provided only by designated national coordinators in FAO member countries. Stock raisers are not directly consulted, even though they may be the only ones who know of many local breeds and strains. To illustrate, only recently did scientists discover a new camel breed with high milk-production potential - although naturally the pastoralists from whom scientists learned of the breed had long known about it.

Neither does the DAD-IS taking account of stock raisers' breed classification systems. These typically differ from scientists' and often are more refined than scientific classifications. To take just one example, scientists opine that India's donkey population is comprised of a single breed. But local donkey-breeders distinguish at least three types that are phenotypic ally distinct and hail from three different areas. By scientific criteria, these features make them, in all probability, three breeds.

The DAD-IS characterizes breeds according to production features and population size. The former include: milk yield, lactation length, milk fat, litter size, birth weight, adult weight and adult wither height. But stock raisers evaluate breeds differently, often using non-productivity-related characteristics. Many such traits are manifest only under traditional management; they may be invisible in stabled animals or on government research farms. For example, transhumant Gujjars of the Himalayan foothills keep a buffalo breed that instinctively knows when to migrate to higher or lower pastures. But this behaviour would be unobservable in Gujjar buffalo penned on a farm or research station. Population data recorded in DAD-IS include, for example, total population size, total number of females bred, total number of males used for breeding, and so forth. Again, this information does not systematically draw on stock-raiser knowledge, breed classifications and terminologies. Consider the case of Tharparkar cattle. Indigenous to the Thar Desert along the Indo-Pakistani border, Tharparkar are famous for their good yields of high-fat milk even under arid conditions. Thus they are accorded a high conservation priority. Yet the word Tharparkar means nothing to local people, who instead say Sindhan. Scientists themselves cannot agree on how to define and hence enumerate this breed. Some count the entire cattle population in the two districts of India where it occurs; others consider only the couple of hundred animals kept on state breeding farms to be 'true' Tharparkar.

ANIMAL DOMESTICATION

Domestication is most commonly portrayed as a fundamental change in the nature of the human-animal relationship. Some authors have envisaged domestication more as a continuation of existing human-animal relations; in other words, as an extension or elaboration of the hunter-prey relationship, rather than as a complete break from it. Domestication can in this way be viewed as the end product of a series of gradually intensifying relationships between humans and other animals. With respect to this, Jarman and Wilkinson (deem it inappropriate to focus merely on the dichotomy between the wild and the domestic.

They suggest that this represents only one aspect of a wide range of close relationships between humans and other animals. Marginal cases, such as the reindeer economies and game-cropping of the present day, they argue, imply that this dichotomy may not have been so clear cut in the distant past. Similarly, Hecker has proposed a broad spectrum of human-animal interactions, ranging from the cooperative driving of animals to specialised hunting and culling to finally the selective breeding of domestic livestock. Although this view of domestication highlights varying degrees of cultural manipulation and control which evolved over time to obtain animal protein, it fails to fully account for the radical changes in human social life and behaviour that animal domestication brought with it in its wake. As Bokonyi points out, although human activities have—in some way or another—always interfered with the lives of other animals, domestication is 'an interference of a quite different kind'.

The profound social significance of the transition from hunting to herding has, however, been incorporated into several recent definitions of domestication. Meadow, for example, describes animal domestication as being a 'selective diachronic process of change in human-animal relationships involving, at the very least, a change in focus on the part of humans from the dead to the living animal and, more particularly, from the dead animal to the principal product of the living animal—its progeny'. This process, he argues, manifests itself in two respects: first, in terms of the transformations in the social and economic structure of the human societies which associate with the animals; and secondly, in the behavioural, morphological and physiological changes which the animal undergoes as a consequence of domestication.

Similarly, Clutton-Brock defines a domesticated animal as 'one that has been bred in captivity for purposes of economic profit to a human community that maintains complete mastery over its breeding, organization of territory, and food supply'. She goes on to argue that domestication is both a cultural and biological process which 'can only take place when tamed animals are incorporated into the social structure of the human group and become objects of ownership'. The morphological transformation of the animal occurs subsequent to its initial integration into human society. The successful domestication of animals was most likely the product of a long-term process of trial and error. It has been suggested that early human efforts to tame other animals were a product of the highly social nature of humankind. Clutton-Brock, for instance, contends that the enfoldment of other species into human society was an extension of the practices of 'sharing, nurturing and protecting weaker members of the human group'. The assumption that our ancestors would have been prepared to tolerate or support weaker persons, other than perhaps infants and young children, let alone members of other species, is somewhat suspect. Historical and anthropological accounts, however, confirm that in some societies, women suckled young mammals along with their human offspring.

This suggests that within early hunting and gathering societies, juvenile animals were perhaps trapped, nurtured and raised alongside humans and were granted a certain level of protection by their human captors. The European explorers of the eighteenth and nineteenth centuries recorded many instances of pet-keeping and affection for small animals amongst the indigenous peoples they encountered. On the basis of their accounts, during the late nineteenth century, Francis Galton postulated that the 'savage' penchant for taming and caring for small animals as pets provided the basis for the development of livestock keeping.

This notion has in fact provided a cornerstone for many explanations of the origins of domestication, most particularly with regard to the early assimilation of dogs into human society. Archaeological findings suggest that *Canis familiaris* was probably the first animal species to undergo domestication. Unlike later domesticates, dogs were not—as far as we are aware—much eaten. Instead, they were most likely used as an aid for obtaining meat. It is widely assumed that the domestic dog descends directly from wolves. Wolves exhibit complex social behavioural patterns similar to humans. They are efficient group predators with a social structure based on a dominance hierarchy. It is thought that our ancestors began to develop a close association with wolves by rearing young cubs that they had caught or found.

The hypothesis follows that some of the more placid of these animals reached maturity and accepted human beings as pack members in their adulthood.

These tamed animals began to breed in human captivity and, over several generations, eventually developed behavioural characteristics distinct from their wild relatives. This process ultimately culminated in the evolution of a separate kind of animal: the dog. Humans and wolves were probably close competitors for food.

They shared the same prey and it is likely that they came into close contact, wolves possibly learning to scavenge on the leftovers from human game drives and the parts of animals which humans preferred not to eat in times of plenty. Although one can speculate on a mutual interest in proximity, it is perhaps more probable that humans saw a way of surpassing their lupine competitors in predation and securing food for themselves by using tamed adult wolves—and eventually dogs—to help in the hunt by detecting and tracking game, and later to help herd other animals rather than prey on them.

The dog is unique amongst early domesticates, given that it was probably not domesticated specifically for food. Most species that underwent domestication were probably intended as 'walking larders'. Sheep, goats and cattle, for example, were most likely exploited as transportable sources of meat and other animal by-products. At a later stage, species such as the horse, donkey and camel were domesticated to provide muscle power for transport and traction, although their meat and milk were also consumed long before they began to perform these roles in human society.

Although humankind was effectively able to secure a constant supply of animal protein for itself through the enfoldment of animals into human society, the fact remains that only a handful of species were ever domesticated successfully.

There are comparatively few domesticated species commonly found in Europe today. The majority of these species originated in the Near East; others descend from the Americas and were only introduced to the European continent following the discovery of the New World. Whilst the appearance and behaviour of these species have changed dramatically since their incorporation into human society, they have flourished under human protection and have, in some instances, outlived their wild or ancestral counterparts. However, as Clutton-Brock points out, the benefits of living under the umbrella of human protection are somewhat dubious. Despite the massive population size and geographic distribution of domesticated animal species in contrast to their wild progenitors, these species have suffered 'irretrievable loss of genetic diversity and evolutionary autonomy'.

DIFFERENT CATEGORIES OF ANIMALS

- *Wild*: Those that breed better where they are free than they do when they are captivated. They have no common use for humans. Example Lion, Tiger, Rhinoceres, Deer etc.
- *Tamed*: Those, which are caught from the wild and trained to be useful to humans in some way. Elephant, Chimpanzee, Gorilla, Yak etc.
- *Domesticated*: Those that are of use at home and are easily bred and looked after by humans. Common domesticated animals are dog, horse, cow, sheep, buffalo, fowl etc.

IMPORTANCE OF DOMESTIC ANIMALS

On the basis of utility, domestic animals are categorised into the following functional groups:

- Milk giving animals:
  - Cattle,
  - Buffalo,
  - Goat,
  - Sheep etc.
- Draught (used for load bearing) animals:
  - Bullock,
  - Horse,
  - Donkey,
  - Mule,
  - Camel,
  - Elephant,
  - Yak etc.
- Fibre, hide and skin yielding:
  - Sheep,
  - Goat,
  - Cattle,
  - Buffalo,
  - Camel etc.
- Meat and egg yielding animals:
  - Fowl (hen) and duck,
  - Goat,
  - Buffalo,
  - Pig etc.

MILK AND MEAT YIELDING ANIMALS

Depending upon the availability and regional considerations different animals are reared for the purposes of yielding milk and meat in India. India is the world's largest producer of milk. The majority of the milk consumed is also in liquid form in India. Over 53 per cent of milk produced in India is from the water buffalo and a majority of milk processing plants in the country depend upon buffalo milk. The *National Dairy Development Board (NDDB)* is the main agency behind the cooperative movement in India. India is now seeking joint ventures and financial participation from the private sector including foreign investment for production of milk and milk products in India.

CATTLE

Cattle mainly include cow, bull, oxen, goat, sheep etc. The females of the species provide milk, which in turn contribute animals protein to the diet of people. While the female species of these cattle are used for milk, the male species play an important role in the agricultural economy by providing labour, meat and hide. Milk itself is taken in many forms like ghee, curd, butter and cheese etc. The excreta of these animals (dung) is used as manure, in biogas and as fuel. There are several important breeds of cattle in India and abroad.

CATTLE FEED

The main feed of cows and buffaloes are grass but this does not provide them all the nourishment. They require balanced diet in the form of roughage which is fibrous food containing large amount of fibres such as hay fodder, leguminous plants-soyabeans, peas and cereals like maize, jowar etc. The diet of cattle mainly consists of roughage (dry or green fodder or fibrous food) and concentrates like grains, oil cakes and seeds, mineral salts and vitamins.

DAIRY PRODUCTS

Milk as drawn from the animals is known as full cream milk. When the cream is separated and the remaining milk is called toned milk. This milk contains no fat and is known as skimmed milk.

On the basis of fat contents the various milk product are as follows:

- *Cream*: It is prepared by churning milk, the fat comes on the top which is separated by draining out the liquid. It is known as cream with 10-70 per cent fat contents.
- *Curd*: Milk is converted to curd due to bacterial activities.
- *Butter Milk*: It is the left over liquid after removal of butter.
- *Ghee*: After heating butter, the water evaporates and fat contents are almost 100 per cent.
- *Condensed Milk*: Milk is concentrated by removing water contents with or without adding sugar. It has 31 per cent milk solids with 9 per cent fats.
- *Powdered Milk*: It is the powdered form of milk.
- *Cheese*: It is coagulated milk protein-casein with fat and water.

- *Khoya*: A desicated milk product prepared by evaporating water contents and reducing the bulk to about 70-75 per cent.
- *Cattle Dung*: Cattle dung is mainly used to make dung cakes for burning as fuels. It is used mainly in villages of India. The farmers also use cattle dung to produce bio gas and the leftover residue as manure.

Biogas Plant (Gobar Gas Plant)

Bio gas plant is a chamber where animal excreta (Cow dung, buffalo dung etc) and some anaerobic bacteria are fed into airtight biogas chamber. Decomposition of excreta produces methane gas used as a smoke free gas for cooking. This gas can also be utilized for lighting. The left over solid residue serves as a good manure.

EGG YIELDING ANIMALS

This category consists of egg producing animals whose eggs are used as food by mankind to provide proteins. Poultry farming is defined as a term for rearing and keeping of birds such as fowl, duck and hen for egg and meat. Poultry farming has become popular because it is comparatively easy to start and maintain. It gives quick return within one to six month of investments,, is easily manageable and requires less space and labour. Poultry birds and their eggs are a rich source of nutrients. Figure gives a comparative account of the composition of fowl meat and that of eggs.

COMMON BREEDS OF POULTRY BIRDS

Indian poultry breeds provide good quality meat but produces small sized eggs. They have natural immunity against common diseases as compared to exotic varieties bred abroad which require greater protection and immunisation.

The chicken is commonly classified on the basis of its origin:

- American,
- Asiatic,
- Mediterranean and
- English
- Plymouth Rock, Rhode Island Red, New Hampshire-American
- Brahma, Cochin, Langshan-Asiatic
- Leg horn, Minoxa-Mediterranean
- Cornish, Australorp-English

INDIGENOUS BREEDS

- *Aseel*: Rajasthan, Andhra Pradesh, Uttar Pradesh
- *Busra*: Gujarat and Maharastra
- *Chittagong*: Eastern India
- *Karaknath*: Madhya Pradesh

Indian Breeds

The Indian breeds of hen include Aseel, Chittagaog, Ghagus and Basra. Their egg laying capacity is around 200 eggs per year.

Exotic Breeds

These breeds are important from other countries and include White leghorn, Minorca, Rhode Island red. These birds have high egg laying capacity but carry less flesh as compared to Indian birds.

Upgraded Variety

Some improved varieties have been developed in India by hybridisation such as B 77, ILS 82 etc. They grow fast and also have as high an egg laying capacity as the exotic varieties and are better suited to the Indian climate.

Poultry Feed

Depending upon the requirement of meat or egg production, poultry feed mainly consists of maize, rice, wheat bran, ground nut cake, fish meal, lime stones, bone meal, common salt, vitamins and minerals.

QUALITY AND SAFETY OF LIVESTOCK PRODUCTS

Quality and safety of livestock products depend upon a quality and safety assurance system for which legislation for setting up standards, corresponding to Codex standards, is obligatory. These do not exist nor is there any method for reviewing and rationalising the quality and safety guidelines. Efforts will also be made for harmonisation of infrastructure facilities for testing food quality and safety with international standards.

DATABASE

Currently, there is absence of a lot of data like those relating to breed-wise milk production of cattle and buffalo, egg production from commercial farms and households, cost of production of milk, egg and wool, availability of livestock resources etc.

A National Animal Health and Production Information System will be established with the active involvement of research Institutions, Government departments, panchayati raj institutions (PRIs), urban local bodies (ULBs), private industries, cooperatives and NGOs. This will work as the national database.

ANIMAL WELFARE

Animal welfare is also related directly with the productivity of animals. The well-being of animals is affected during management under the intensive production system, in the animal market, during handling and transportation, rearing of buffalo male calves in urban areas etc. There is a great deal of wastage, as well as animal suffering due to ill-designed agri-implements, carts and implements attached to animals.

Efforts will be made to strengthen the institutions working on a livestock care system so that they can ensure and promote animal care and well-being. Research and technology development will be taken up for enhancing efficiency and reducing drudgery of animals by improving the design of carts, yokes, implements and toolbars used in agriculture. A good example is the buffalo-drawn bogey fitted with rubber tyre and bearings.

DEVELOPMENT OF LOCATION SPECIFIC ANIMALS

Camel will continue to be important in desert areas for quite some time. Effective support for providing nutrition and health cover is needed for its improvement. The Department of Animal Husbandry will continue its programme for improvement of better studs both for horses and donkeys used for transport in hilly areas.

Horse riding is now becoming an integral part of amusement parks and this will be encouraged as a niche industry. To encourage the breeding of horses, mules and asses, technological and financial support will be extended to entrepreneurs. Animals indigenous to specific agro-climate regions like Yak and Mithun will be developed.

CAPITAL FORMATION

Public sector lending in the livestock sector is low and inadequate credit support leads to poor capital formation. As the organised financial sector is unwilling to finance livestock programmes that are not in their interest, especially after the initiation of financial sector reforms, the livestock farmers are mainly dependent on the financial intermediaries and they end up bearing a higher interest rate than would be available otherwise.

Attempts would be made to create a favourable economic environment for increasing capital formation and private investment. Financial institutions would actively participate in livestock credit programmes through standardised ready-made bankable projects with back-ended subsidy. Creation of a venture capital fund is needed to assist the private entrepreneur in establishing units that could provide services and goods at the district/block level.

NEW FRONTIERS OF ANIMAL USE

The scientific developments of recent years often seem like the stuff of science fiction, rather than science fact. Almost daily one can leaf through the newspapers, or turn on the television, to be confronted with news of the latest advances that will purportedly provide new hopes for the treatment of disease or the manufacture of food. Whether it be for the production of a cheaper and more effective vaccine to treat hepatitis B, or for the manufacture of a tomato that contains fish genes so that it can be successfully frozen, 'genetic engineering'—as it has been popularly dubbed—makes for good headline news because it is a topic that captures the public and scientific imagination alike. Genetics has been heralded as the science that will transform agriculture, cure diseases and conserve other species, although it could potentially also be employed to more sinister and destructive ends.

Essentially, molecular genetics is a science that explores the mechanisms of heredity. Each individual, human or animal, possesses a unique genetic 'blueprint' that makes him or her different from any other living creature. This genetic blueprint can, however, be modified. This can be achieved through a recombinant DNA technique that involves the injection of a foreign gene into an organism in order to produce specific new characteristics that in turn will also go on to be inherited by future offspring. Organisms that have been modified in such a fashion are known as 'transgenic organisms'.

The science of molecular genetics has found an increasing number of applications with regard to food-producing animals and veterinary science. In recent years, the gene mapping of farm animals has become an important tool for breeding and has been employed to localise, isolate and characterise the genes that are responsible for specific traits important to both health and production. With the development of new molecular genetic techniques that will help to isolate and identify the DNA markers that are linked to the genes responsible for economically important production traits and disease resistance, animal breeders will in the future be better equipped to single out the animals carrying the most desirable genes.

Theoretically, once such genes are identified, animals may be genetically selected or engineered for disease resistance. Another related application of genetic engineering is the production of more effective medical and veterinary vaccines. This may offer livestock producers a viable solution to the serious problem of enhanced infection risks that has been caused by increasingly high stocking densities. Advances in DNA recombinant technology have created the possibility for live carrier virus vaccines' to be developed that can be tailored to veterinary needs. It is thought that such modified live viruses will be far cheaper to manufacture and more efficient in the prevention of animal disease than existing ones.

Likewise, biotechnology—a term that may be generally applied to the manipulation of organisms for commercial purposes—has found important uses within animal reproduction. For many years, artificial insemination has provided the basis for the systematic genetic improvement of animals. A more recently developed reproductive technique, *in vitro* fertilisation (IVF), may possibly be employed within livestock production in the years to come. IVF involves the maturation of the egg outside the body and its fertilisation in a test tube. The embryo is then implanted into the womb of a surrogate mother. Although IVF in cattle is still in its infancy, embryo transfer is a technique that is now widely employed in livestock species and has various advantages with respect to genetic improvement, health protection and productivity.

Biotechnology has also been applied to improve animal nutrition. It has been utilised at three levels: in the feed, in the digestive tract (*i.e.* in relation to the ruminal microflora that aids digestion) and in the physiological regulation of the animal. Genetic technology can, for example, be employed to alter rumen bacteria in order to manipulate the animal's digestive tract ecosystem. Thus far attempts to interfere with animal metabolism have met with considerable resistance. In this respect, it is the application of an active agent, the growth hormone bovine somatotrophin (BST), to increase milk yield in cattle that has caused perhaps the greatest consternation, raising serious questions about the lengths to which livestock producers may be willing to go in order to increase productivity. As mentioned in Chapter 5, the use of BST, which is currently produced from transgenic microbes, has been prohibited within the European Union.

One of the most controversial applications of recombinant DNA technology has been the creation of transgenic animals. The first experiments with transgenesis in mammalian species took place with animals used for laboratory experimentation. For instance, during the mid-1980s, a transgenic mouse, christened the 'oncomouse', was developed by biomedical researchers at Harvard University in their bid to understand and find a cure for breast cancer. The 'inventors' of such transgenic creatures have even gone so far as to patent their creations, leading to heated debates upon the nature of intellectual property rights and their application to living organisms.

Not surprisingly, such genetic technology has also found its way into the realms of livestock production. Since the mid-1980s, fledgling experiments with transgenesis have taken place involving food-producing animals, such as sheep, goats, pigs and cattle. In the first instance, the creation of transgenic farm animals has been seen as a possible means to increase and improve agricultural productivity, to meet increasing human demands for animal produce without incurring further detriment to the environment by requisitioning extra land for agricultural production.

However, some of these transgenic experiments with food-producing species have had quite different aims. One of the prime incentives to create transgenic farm animals has been to produce substances known as biopharmaceuticals in milk. Bio-pharmaceuticals are substances produced by human genes that are essential to fight off disease and keep healthy. When such genes are incorporated into the DNA of other organisms, *e.g.* bacteria or yeast, these organisms become able to produce biopharmaceuticals beneficial to human health upon a scale that is commercially viable.

A prime example of such experimentation with transgenesis and the production of biopharmaceuticals can be provided by Herman, the world's very first transgenic bull. Born in 1990-and the result of a unique collaboration between the biotechnological company Pharming BV and a Dutch governmental research institute—Herman was the product of a fertilised egg in which the DNA had been modified to include a synthetic gene identical to the human one that controls the production of lactoferrin, a milk protein that has strong anti-microbial properties and stimulates intestinal flora to provide protection against bacterial infections. Apart from this single gene, Herman is essentially no different to any other bull with respect to his genetic make-up. Herman was created specifically in order to sire female offspring who would inherit this foreign gene and consequently produce human lactoferrin in their milk.

The first mature milk from Herman's transgenic progeny became available early in 1996 and was found to contain the appropriate lactoferrin. The commercial implications of such successful biotechnological developments are potentially enormous, for they offer new possibilities for the production of pharmaceutical drugs to treat human conditions at a far lower cost. For example, Pharming, the biotech company responsible for Herman and his daughters, is currently attempting to develop a range of human health care proteins in transgenic cattle.

The human lactoferrin that they have developed is intended as a component in clinical nutrition that can be used in the treatment and prevention of bacterial gastrointestinal infections in patients who have been immuno-compromised. A further planned application is in the production of speciality medical formulas for newborn and premature babies. However, perhaps more importantly, the introduction of this new transgenic animal technology signifies a new role for domesticated animals within human society. It is likely that in the not-too-distant future, animals will not only continue to function as the providers of protein, but may also become living, breathing and walking 'pharmaceutical factories'.

Transgenic technology may also find other important human health applications in future years, particularly with respect to the use of animal organs for transplantation into human beings. Since organ transplant technology was first developed for the kidneys and liver and then first employed in human heart surgery by Christiaan Barnard in 1967, organ transplantation has become increasingly important in the treatment of life-threatening conditions. The continued success of surgical procedures such as kidney and heart transplantation has led to an ever-increasing demand for healthy human organs, yet—unfortunately for those awaiting such radical, though potentially life-saving, operations—there is a great shortage of suitable organs. As the population ages, it is inevitable that this demand will increase further still. In spite of numerous high profile campaigns to encourage people to carry donor codicils, the shortage of donor organs has remained critical, inspiring scientists to search for new alternatives to human organs for transplantation. Some have sought solutions in the form of artificial organs; others have begun to seriously investigate the possibilities of using animal organs for transplantation into humans.

Xenotransplantation, as such cross-species organ transplantation is known, has been heralded as a potential solution to the organ crisis, for it could provide a ready-made supply of healthy donor organs to the thousands of people who today, and in the future, will require transplant surgery. Because of the animals' genetic affinity to humans, scientists initially turned to primates—particularly baboons—in their attempts to transplant organs to humans. The very first attempt at clinical xenotransplantation in fact occurred as early as 1910 when a German medical scientist, Dr Unger, transplanted a monkey kidney into a uraemic girl.

More recently, in 1992, a baboon liver was transplanted into an HIV patient who was dying of hepatitis B in the USA. Primates, however, pose particularly high risks to humans with regard to the transmission of zoonotic disease, raising serious questions as to their suitability for use in xenotransplantation. The recent emergence of new highly infectious diseases such as Ebola and Marburg have further emphasised the potential role of primates as the reservoirs of viral diseases deadly to humanity. Further to this, the use of primates in medical experimentation has become increasingly controversial due to a growing sensitivity to the semblance between humans and primate species. In recent years, calls have even been made for the great apes to be accorded equal rights to humans.

In view of such developments, scientific attention has instead turned to the domestic pig in the bid to produce organs suitable for human transplantation. Although more distantly related to people, the pig has already proved to be the ideal provider of insulin to treat human (and pet animal) diabetes and valves to correct heart abnormalities. Porcine organs are also approximately the same size as human ones. Moreover, given that these animals breed most prolifically, they could potentially provide a regular and reliable supply of organs for transplantation. All this being said, there are major problems that would have to be overcome before pig organs could actually be transplanted into the human body. Organ rejection is probably the greatest obstacle. This commonly occurs in the transplantation of human donor organs and is generally counteracted by the life-long usage of anti-rejection drugs by the organ recipient; in xenotransplantation rejection is likely to be hyperacute.

One of the solutions to this problem that is currently being sought by scientists is to employ transgenesis in order to genetically modify the porcine donor's organ in a way that will trick the human recipient's body into accepting it. The hearts of transgenic pigs that have been bred by researchers have already been transplanted into monkeys with some degree of success, arguably giving some indication of future possibilities for human application. The question of disease transmission through such transplant technology is also here at issue.

Some porcine pathogens may indeed pose risks to human health, though it is more than likely that pigs intended for xeno-transplantation will be bred to be specific-pathogen free. In fact, some may even be inclined to argue that the option of transplanting animal organs carries *less* disease risk than the transplantation of human donor ones. The animals used would necessarily have been genetically modified, bred and kept under strict clinical conditions; unlike the human donor who may carry all manner of diseases that may go on to infect the recipient once the organ has been transplanted.

Another recent development in the science and technology of genetics that may find its most ideal application with regard to producing transgenic animals for both transplantation and biopharmaceutical purposes is cloning. Once relegated to the realms of science fiction, cloning has today become very much a reality. In February 1996, the public and scientific world were stunned by the announcement that two genetically identical lambs, named Megan and Morag, had been born at the Roslin Institute in Scotland. These sheep had been cloned by the process of nuclear transfer from a cultured cell line, originating from different cells of the same embryo. Morag and Megan provided concrete proof of the possibility of creating animal life without the need for male sperm

In February 1997, the even more astonishing news broke that a lamb had been cloned—by the very same scientific team—from the cells of an adult sheep. To create Dolly, as she was christened, DNA from a single cell had been taken from another sheep's udder, making her the very first mammal to have been created from the non-reproductive tissues of an adult animal. Dolly the sheep became a celebrity overnight and her very existence precipitated fervent debates about the ethics of cloning and its potential application to humans. A month later, hot on the hooves of this story, came the news that Australian researchers had managed to create almost five hundred genetically identical cattle clones. These were created from egg cells that had been taken from cow ovaries. At the time of writing, it remains to be seen whether these embryos can be advanced through successful pregnancies to birth.

The potential consequences of cloning are enormous. In the future it may well be possible to produce hundreds of copies of an adult animal. Such cloning would certainly assist the production of biopharmaceuticals in milk. Moreover, it could be feasibly used to create animals that grew faster for meat production and that were more resistant to disease; though one may imagine that a herd of identical animals may indeed be at a far greater risk of disease. Presumably, cloning could also play a role in protecting the bio-diversity of species by ensuring that breeds would not die out through keeping the appropriate cells and embryos in cold storage; though at the same time it poses an enormous threat to the genetic diversity of animal species.

From the point of view of both the consumer and livestock producer, cloning could theoretically ensure that meat and milk of a reliable and standard quality could be produced and sold at a standard price and would turn a standard profit. In this sense, cloning is an extremely attractive proposition for the livestock industry; as is transgenesis also—presumably these technologies will also be combined. They are, however, scientific phenomena with which many already feel extremely uncomfortable. Such technology necessarily involves direct human interference in—what is often perceived to be—the natural order and raises serious questions about the ethics and morality of such scientific enterprise.

At present, national governments and scientific regulatory bodies have chosen to err very much on the side of caution when it comes to biotechnological experimentation. Often it is only when it can be proved that genetic experiments with animals can provide important benefits to humankind—as has been successfully argued for the transgenic production of lactoferrin—that they are permitted under stringent controls. Legal prohibitions on the application of such genetic technology in human beings are already firmly in place, particularly with regard to the cloning of human embryos. It remains to be seen whether either animal cloning or the production of transgenic animals will gain widespread acceptance or become commonplace in the new millennium.

Clearly, the scientific developments of recent years and the impending future present a colossal moral quagmire through which scientists, law makers and ethicists alike will be forced to wade; undoubtedly, there are also likely to be significant implications for animal welfare. However, there are also a whole host of practical problems that would have to be overcome before such technology could be successfully and safely applied. The practice of genetic engineering, for instance, carries with it an inherent risk because of pleiotropy; that is a phenomenon which occurs when a gene that does one specific thing may have other, possibly unrelated, effects upon or could interfere with the behaviour of other genes.

One of the major and commonly espoused objections to the genetic technology that has been developed in recent years is that it necessarily involves the deliberate and wholesale interference with 'nature's design'. Further to this, the idea of genetically modifying other life forms has also met with considerable hostility for it reeks of eugenics; the implications of which are still fresh in our collective memory from the darker events of the twentieth century. History has already taught us that the calculated 'improvement' of humans can be a highly dangerous business with the most sinister of consequences. Yet leaving the issue of the application of genetic technology to humans aside and focusing upon animals alone, it is pertinent to consider just how well grounded our objections to and fears of the genetic modification of animals actually are.

Although the genetic technology of recent times has often been portrayed as an extremely new and outlandish development, the fact remains that there is—at least in principle—very little new about it; most certainly as far as animals are concerned. The genetic modification of animals can logically be viewed as an extension; or rather a continuation of the process of domestication and selective breeding that has already been going on for the past 10,000 or so years. As this book has sought to illustrate, for thousands of years, humankind has actively attempted—in ever-increasingly successful measures—to alter and improve the behaviour, physical appearance and productivity of animals in order to service its needs.

Whether it be to produce a more tractable animal or one that will produce more milk or larger calves, animals have persistently been deliberately and selectively bred to achieve highly specific ends. In this sense, people have been practising the art of 'genetic engineering' since animals were first domesticated. Traditionally, new varieties of animal species have been produced by cross-breeding. Until very recently, breeders had no detailed knowledge of the genetic mechanisms involved in producing such new varieties. Animal husbandry was a purely empirically based occupation. Farmers gradually learned that if, for example, cattle that produced a high milk yield were bred with other high-milk-yielding cattle, they would produce offspring that would go on to produce even more milk.

This kind of genetic manipulation is, however, far from an exact science. Moreover, such selective breeding can take an extremely long time before it produces the desirable result. Recombinant DNA technology has expedited this process, for it has provided the means by which the genetic make-up of animals can be modified in order to produce specific traits that—by traditional means—would have taken many generations to introduce, or would never have occurred at all since they involve the insertion of DNA foreign to the recipient animal.

Likewise, although many tend to balk at the mention of it, there is little new about biotechnology. Biotechnology can essentially be defined as 'the application of scientific and engineering principles to the processing of materials by biological agents to produce goods and services'. Whilst this may sound like a new-fangled idea, the fact remains that humankind has been using biotechnological techniques for many thousands of years in order to manufacture products such as beer, wine and bread.

To make such products, microorganisms have been routinely and deliberately manipulated for highly specific ends. The role of yeast in bread baking is perhaps the best example of this. Again, it has been the development of recombinant DNA technology that has increased the capabilities of biotechnological production, rendering it an important and efficient technique for the future. As has already been suggested, it is likely that biotechnology will play an increasingly important role in veterinary medical science, particularly with respect to the improvement of animal health and nutrition, the production of veterinary vaccines and prevention of animal disease. As a corollary to this, biotechnology may also be crucial to the maintenance of public health as it may contribute greatly to the control of zoonotic disease.

There are, however, good reasons why we should be cautious about the application of recombinant DNA technology to domesticated animals. Selective breeding, in the traditional fashion, has already proved to have had serious consequences for animal health and welfare. Take, for example, the current state of broiler chickens and turkeys that have been bred to grow so fast and produce such grossly enlarged breast muscle tissue that their legs become crippled; or the double-muscled Belgian Blue calves that necessarily have to be born by caesarean section. Such breeding practices have been developed in the constant bid to increase productivity to meet consumer demands and to produce higher profits, generally at the expense of animal well-being. But, as Chapter 6 illustrated, it is not only the food-producing species that have fallen foul to irresponsible breeding practices.

The health and well-being of pet animals have also been compromised severely through breeders' continued attempts to produce new varieties and to emphasise specific animal characteristics. One could argue that the application of genetic technology could plausibly solve many of the problems created by both intensive farming practices and pet-keeping. For example, it may in the future be possible to create a race of hornless cattle, thus obviating the need for dehorning altogether. Likewise, transgenic chickens might be produced that lack nerves to their beaks and could thus be debeaked without causing them any pain.

Whilst such developments may indeed be possible, they will nonetheless do nothing to improve the general conditions in which intensively farmed livestock live and are bred. Similarly, transgenic cats could be produced without claws and dogs with modified vocal cords for the convenience of their human owners.

It remains to be seen whether the genetic modification of animals will create serious new problems with regard to animal welfare. The application of recombinant DNA technology to animals is so recent that it cannot yet be said with any great certainty whether, for example, the inclusion of a single human gene to cattle in order to produce biopharmaceuticals will endanger animal health and well-being in any fashion. Nor for that matter whether it may affect the health of humans who consume such genetically modified milk or its pharmaceutical derivatives. It is likely that such cattle will essentially continue to function just as and be treated like any other dairy animal. In that sense, transgenesis may change little with regard to how animals will live within human society.

Creating transgenic pigs (and possibly other species also) for xenotransplantation, on the other hand, will most likely entail that the animals be kept in extremely sterile, clinical environments that are a far cry from the conditions, for better or worse, that they now experience. What is, however, clear is that with the introduction of genetic engineering, our manipulation and exploitation of animals has entered a brand new phase. The production of new animal varieties can now take place over the course of one single generation, rather than many. Moreover, and perhaps more importantly, recombinant DNA technology means that animals can be modified using genetic material that is completely alien to the recipient animal species. Needless to say, this signifies a crucial new development in the age-old practice of animal husbandry. It remains to be seen to what extent the practice of genetic engineering will be continued and commercially employed, or whether this will be deemed ethically and biologically unacceptable.

As the new millennium approaches, it is clear that the role of domesticated animals in human society is continuing to diversify, rather than diminish. Whilst it appears that many people are becoming more sensitive to the lot and rights of other animals, humanity as a whole seems to be perpetuating the tacit belief that animals may be exploited as a natural resource. In this respect, the message of this book is rather bleak for those who seek to reduce or rid the world of animal exploitation. For the foreseeable future, this is unlikely to happen; though it does not necessarily exclude the possibility that animals could well be treated in a more humane fashion whilst being used by humankind.

A world where animal flesh is no longer eaten, nor animals exploited for human ends in any way, remains steadfastly within the realms of fantasy and science fiction. The Utopian vision of *Star Trek* where humanity has attained the pinnacle of civilization by no longer breeding and killing animals for food, instead obtaining its nutrients and raw materials through synthetic replication, is far removed from the stark realities of the twentieth—and probably twenty-first—century. Contrary to what many may hope, the human exploitation of animals has greatly increased and looks set to increase still further in the years to come as new uses for animals are dreamt up along with the new technologies that can achieve them.

The final question remains of just what these developments will entail for the veterinary regime. It seems likely that as the frontiers of animal exploitation continue to expand, so will the importance of veterinary medical science to society also. Veterinary expertise has thus far been crucial to the development of transgenic animals for the production of biopharmaceuticals and xenotransplantation. The close interaction between veterinarians and scientists from other academic spheres will presumably continue as such technologies are further refined or expanded. Scientific developments seem likely to require that in the future veterinarians will be obliged to be complicit in the activities of the biotechnologists, as much as they have become in the activities of the livestock industry during the post-war period.

If biotechnology turns out to offer all that it today seems to promise, then it seems certain that the efficacy of the veterinary regime will improve further still. The prospect of new and more effective vaccines to combat animal disease and the improvement of animal nutrition have great implications for both the health and well-being of animals and public health. If the damage caused by animal disease is reduced and the methods of controlling it become cheaper, the consumer will undoubtedly also reap the benefits of veterinary advances through being able to buy safer, cheaper and better quality animal produce. However, at present, many consumers are skeptical about the advantages that genetic technology purportedly offers and the impact that it will have upon their lives.

Likewise, policy-makers and legislators have responded cautiously to recent scientific advances and have been keen to exercise their regulatory powers upon the burgeoning biotechnological industry. In this respect, as we enter the twentyfirst century, the state is likely to tighten its grip still further and continue its imposition of the veterinary regime with increased vigour, particularly within an international framework. Throughout the course of the late twentieth century, the importance of an effective veterinary regime has become increasingly evident. Calamities such as the recent BSE crisis have explicitly highlighted the potential consequences of our increased exploitation of animals and the necessity of intensifying measures for animal disease control. As the new millennium dawns, it seems more than likely the burden of beasts for human society will continue to be a heavy one.

NEW HEALTH PRODUCTS FOR FARM ANIMALS

Animal diseases cost U.S. agriculture $17 billion each year, according to the Congress Office of Technology Assessment. To help prevent these losses, animal scientists are using biotechnology to develop an array of products to diagnose, treat, and prevent disease in farm animals. An area where animal biotechnology has already had a profound impact is diagnostic testing. Many animal diseases are difficult to diagnose. A veterinarian often has to wait hours or days for laboratory results to confirm a diagnosis. In the meantime, the veterinarian must either withhold treatment or risk using an inappropriate therapy. Using biotechnology, scientists are developing fast, accurate diagnostic tests for many of the most common farm animal diseases. In some cases, a veterinarian can conduct the diagnostic tests right on the farm and immediately start the best therapy.

Many of these new tests use monoclonal antibodies. Antibodies are proteins an animal's immune system produces in response to invasion by bacteria, viruses, or parasites. Each kind of antibody binds specifically with a characteristic part of a particular kind of invader. Scientists make monoclonal antibodies by fusing two kinds of cells. One is an immune system cell that produces antibodies, which bind to part of a particular disease-causing microbe; the other is a cancer cell. The cell that results from the fusion inherits from the immune system cell the ability to produce antibodies, and from the cancer cell the ability to reproduce indefinitely.

This new hybrid cell, or hybridoma, acts as an antibody factory. The antibodies are called monoclonal because they are produced by identical "clones" of the original hybridoma-all will bind with the identifying structure on the surface of the microbe in question. One application of monoclonal antibodies is as a test for brucellosis in cattle. This bacterial disease often causes cows to abort pregnancies, and it can be spread to farmers and people who drink milk from infected cows.

Vaccines protect pregnant animals from abortion, but vaccinated cows may become carriers of the disease. Conventional diagnostic tests cannot distinguish between the disease-causing microbe and the vaccine, which is made from the microbe. Thus, these tests cannot help the farmer determine which animals are carriers of the disease and which are not. But diagnostic tests using monoclonal antibodies are so specific, they distinguish between cattle that carry the disease-causing bacterium and those that carry only the vaccine. With the aid of these new tests, farmers can isolate carriers to prevent the spread of infection. Diagnostic tests are not just for diseases.

A monoclonal antibody test to detect pregnancy in animals is so simple that a farmer can read the test right on the farm. Monoclonal antibodies are also used in a test to detect a disease called scours in piglets and calves. Scours is a bacterial disease of newborn animals that causes diarrhea. Many of the afflicted animals die of dehydration.

By providing quick, accurate diagnosis, the monoclonal antibody test for scours permits immediate administration of therapy. The therapy for scours in calves is also based on monoclonal antibodies. Infected calves are fed monoclonal antibodies that coat the offending bacteria, preventing the microbes from causing diarrhea. Some other therapies for animal diseases are based on recombinant DNA technology. Genetic engineering allows scientists to insert one or more new genes into an animal, plant, or microbial cell. In developing treatments for an animal disease, for instance, an animal gene is inserted into a microbe or an animal cell that can be grown in a laboratory culture. The inserted gene is one that produces a natural disease-fighting protein. The microbes or animal cells become minute factories, producing large quantities of the therapeutic protein. For example, genetic engineering is used to produce interferons and interleukin-2, natural proteins that fight viruses.

Veterinarians long have used antibiotics against bacterial diseases, but they have had few drugs to treat viral infections. Interferons and interleukin-2 not only kill viruses directly, they also stimulate the animal's immune system, improving the immune system's ability to fight disease naturally. One target of research regarding interferons and interleukin-2 is shipping fever, a disease of cattle caused by several viruses infecting an animal simultaneously. Although the animal's immune system normally keeps the viruses in check, outbreaks of the disease may result from stress during transport to feedlots or market. Shipping fever costs the beef industry more than $250 million a year. Injections of interferons or interleukin-2 before shipping may help to augment the cattle's natural immunity and prevent shipping fever.

BETTER VACCINES TO PREVENT DISEASE

In addition to producing new diagnostic tests and therapeutic proteins, animal scientists are using biotechnology to develop vaccines to prevent disease. A safe, effective vaccine for swine pseudorabies, a fatal herpes-virus, is already in use. An advantage of the swine pseudorabies vaccine, and other vaccines made through recombinant DNA technology, is that they use only a small portion of the original microbe. Conventional vaccines use killed or weakened forms of the disease-causing microbe.

Sometimes, when the microbes are not completely killed or sufficiently weakened, a conventional vaccine may cause the disease it was supposed to prevent. Disease-causing genes are not included in the genetically engineered vaccines. Therefore, the recombinant vaccines build up the body's immunity without the risk of causing disease. Another advantage of recombinant vaccines is the speed with which they can be developed. "Conventional vaccine development can take 20 or 30 years, maybe even 100 years," says Karen Jacobsen, D.V.M., University of Georgia College of Veterinary Medicine.

"This is one reason why we don't have vaccines for a lot of animal diseases today". Recombinant vaccines are being developed for foot-and-mouth disease, a highly contagious viral disease that infects cattle, sheep, and other animals. Although this disease has been eradicated in North America, it still causes substantial livestock productivity losses elsewhere, particularly in developing countries. Infected herds must be slaughtered, and contaminated ranches must be left idle for months to prevent new outbreaks of the disease. Conventional foot-and-mouth vaccines are made by weakening the virus that causes the disease.

These vaccines sometimes revert to the virulent state, and they have caused outbreaks of the disease in Europe. The foot-and-mouth vaccine made with biotechnology cannot cause the disease because, as in other biotech vaccines, the disease-causing genes have been removed. The biotech vaccine does not need to be kept cold and, therefore, can be used in developing countries. In addition to being used in treating swine pseudorabies and foot-and-mouth disease, recombinant DNA technology is being used to develop vaccines against dozens of livestock and poultry diseases, such as scours and a variety of respiratory disorders. Some vaccines may be injected into eggs so that chickens are immunized before they hatch. Most vaccines protect against viral or bacterial infections, but using genetic engineering, researchers in New Zealand and Australia have developed a vaccine against a parasitic disease. Farmers must destroy the meat from sheep infested with a species of tapeworm common in those two countries. The new vaccine, which protects sheep from the tapeworm, may save farmers millions of dollars annually.

PROTEINS FOR ANIMAL NUTRITION

Natural proteins called somatotropins, or growth hormones, help animals convert feed into muscle or milk. Using recombinant DNA technology, scientists have been able to develop bacteria that produce commercial quantities of somatotropins.

They have found that administering small additional amounts of these proteins to animals helps to increase the animals' efficiency of feed conversion, thereby saving the farmer money. Porcine somatotropin (PST) may help improve human health, as well as lower the farmer's cost of production. PST improves feed efficiency in hogs by 15 to 20 per cent. It also reduces fat deposition, allowing PST-treated hogs to provide consumers with leaner cuts of pork. The ability to produce lean pork has enormous implications for improving human health by reducing dietary fat and cholesterol. Another growth protein, bovine somatotropin (BST), is being administered to make dairy cows more efficient, thereby lowering the dairy farmer's costs. Regular administration of small amounts of BST increases a cow's milk yield by 10 to 25 per cent. Although feed intake also increases, this feed increase is proportionately less than the increase in milk production.

Studies indicate that BST can improve the milk-to-feed ratio by 5 to 15 per cent. Extensive laboratory testing has shown that the meat and milk of treated animals contain no more PST or BST than the meat or milk from untreated animals. Other tests have shown that these proteins do not act on the human body, and therefore, they are safe to eat. Nevertheless, the use of BST remains controversial. Testing is under way to make sure that animals are not adversely affected by treatment with somatotropins. Another protein that can be administered to animals to increase feed efficiency is called growth hormone releasing factor (GHRF). While not a growth protein itself, GHRF causes the animal to increase production of growth proteins.

IMPROVED ANIMAL BREEDS

Long before anyone knew about genes or DNA, people were selectively breeding animals to produce desirable traits, such as strength, feed efficiency, and disease resistance. Selective breeding is a crude technology, because each cross between two animals mixes hundreds of thousands of genes, both desirable and undesirable. Recombinant DNA technology is now making it possible for scientists to breed animals with great precision.

One or more new genes can be inserted into an animal embryo without disturbing the rest of the animal's genes. The resulting animal with the new gene or genes is referred to as a transgenic animal. Despite the differences in technique, the scientist using recombinant DNA has the same objectives as the selective breeder: improving efficiency, disease resistance, and other desirable traits. But recombinant DNA provides a wider array of possible new traits, since recombinant techniques allow animal scientists to insert genes not only from the same or closely related species, but also from distantly related animals, plants, and even microbes.

One possible application of gene transfer in animals is to develop dairy cows that produce more nutritious milk. Bottle feeding human infants with cow's milk is not an ideal substitute for breast feeding, because cow's milk differs from human milk in some important components. Scientists are looking for ways to transfer genes into dairy cows that will enable the cows to produce milk that more closely resembles human milk. Advances in Control of Parasitic Diseases Continue to Rely on Biotechnology Parasitic disease continues to be a major health problem worldwide in humans and domestic animals. Applying biotechnology has led to progress in prevention and control in many areas, particularly in regard to diagnosis.

Parasites, like most organisms, carry repetitive sequences in their genomes. These are sometimes tandem repeats (back-to-back in a chromosome), or they may be located at random throughout the genome. Often, repetitive sequences are intergenic and non-coding, and are thus far more subject to evolution and change, increasing the likelihood that they will be genus or even species specific. Since detection of these sequences is not dependent upon gene expression, which may vary over time, repetitive sequence-based assays allow detection of the parasite at any life stage.

The effectiveness of repetitive sequence-based polymerase chain reaction (PCR) methods of identification has been demonstrated for parasites of several clinically important genera (1). A PCR assay has also been developed to differentiate two closely related nematode species that are virtually indistinguishable by morphologic examination (2), as well as for diagnosis of Toxoplasma and Cryptosporidium infections (3). Attempts to develop anti-parasite vaccines have met with multiple failures. The problems often result from lack of post-translational modification by currently used vector-host systems and the inability to regenerate the secondary and tertiary structure of the native immunogens. Antigen presentation in recombinant parasite vaccines has also been thought to be a problem resulting in failed efficacy. There is now some evidence to indicate that cloning of genes for antigens of interest to yield chimeric molecules expressed in the BCG strain of Mycobacterium bovis may provide improved antigen presentation to stimulate both humoural and cellular responses (4); however, it is unlikely that future applications of this technology will be simple. Nonetheless, work is proceeding on the application of biotechnology to the development of vaccines against parasites in most food animals.

A particularly striking exception to the marginal successes reported in other systems has been a recombinant vaccine for a parasitic helminth of sheep (5). This product yielded greater than 94 per cent protection in experimental work and is now available as a commercial product. A further possibility is the application of anti-idiotypic monoclonal antibodies. When monoclonal antibodies are prepared against a monoclonal antibody, it is possible to obtain, in a very small number of cases, an antibody which mimics the original antigen. Results of work with Cryptosporidium perfringens epsilon toxin suggest that animals immunized with anti-idiotype monoclonals can be protected against toxin challenge.

7

Dairy Cattle

Dairy cattle include replacement heifer calves and yearlings, dry cows, lactating cows, and breeding bulls used for research and teaching purposes related to milk production. The basic requirements for safeguarding the welfare of dairy cattle are an appropriate husbandry system that meets all essential needs of the animals, and high standards of handling.

MILKING MACHINE AND UDDER SANITATION

The milking facility should have a programme for regular maintenance of milking machines and follow the recommended mastitis control programme of the National Mastitis Council. Appropriate equipment and competent personnel should be available for milking. Personnel responsible for milking should receive ongoing training about proper milking procedures as the frequency of training has been associated with adequacy of milking performance.

Animal care facilities should be designed and operated to standards meeting or exceeding those of grade A dairies as defined in the Pasteurized Milk Ordinance. Areas where milking takes place must be designed and constructed in accordance with the 3-A Sanitary Standards Inc. Accepted Practices. Cows should be maintained in housing areas that provide for adequate hygiene to ensure that udders are visibly clean. Cows should be milked on a regular schedule that is appropriate for the goals of the herd or specific research project.

Written operating procedures should be established to control potential contamination of milk with antibiotics or other pharmaceutical agents. Antimicrobial treatments should be administered based on approved defined protocols. All extra-label treatments must be administered under the supervision of a veterinarian that has an appropriate veterinary-client-parent relationship.

Milking machine and udder sanitation are vital to an effective preventive programme against mastitis and follow guidelines as established by the NMC. Care should be used to minimize the excessive use of water before and during udder preparation. Emphasis should be placed on ensuring that cows enter the milking parlor with clean, dry teats. Udders, especially teat ends, should be clean and dry when teat cups are applied for milking. The removal of foremilk before teat disinfection is encouraged as a means to detect mild cases of clinical mastitis. Teat sanitation, predipping, and wiping immediately before machine attachment reduce udder infection caused by environmental pathogens. Postmilking disinfection of teats is an essential management practice that greatly reduces the incidence of mastitis. Milkers handling cows should pay meticulous attention to their own personal hygiene and wash their hands thoroughly before milking and frequently during milking.

The use of clean nitrile or latex gloves during milking is highly encouraged to prevent contamination of the udder. Cows with subclinical cases of contagious mastitis should be milked last to reduce the spread of mastitis throughout the herd. Udder hair removal is recommended as a means to improve milking hygiene and udder health. Cleaning of milk handling equipment is accomplished by a combination of chemical, thermal and physical processes and cleaning regimens should be designed to meet appropriate regulatory standards.

Recommended cleaning and sanitizing practices are a balance between the cleaning temperatures, cleaning chemical concentration, contact time and mechanical action. Effective cleaning programmes for milking machines include use of hot water; use of disinfectant solutions and other chemical agents effective for removing mineral, milk fat, and protein deposits from equipment between milkings; disinfection of teat cups between cows; and flushing of teat cups with warm water, cold water, boiling water, or chemical disinfectant solution.

The most common routine in the United States is a combination of prerinse, alkaline detergent, acid rinse and premilking sanitize. Very small herds may utilize manual cleaning and disinfecting that involves hand-cleaning of some or all of the milk harvesting and storage equipment. Small to medium herds commonly use automatic washing equipment.

This equipment will automatically mix the chemicals with the appropriate water volume and temperature and circulate these solutions through the milking machine. On large farms, an attendant may be present to mix chemical solutions and operate valves for circulation. The effectiveness of milking system cleaning can be evaluated by examination of standard plate counts and laboratory pasteurized counts performed on bulk tank milk samples.

STRAY VOLTAGE

The term stray voltage describes a special case of voltage that develops on grounded metal objects on farms. If this voltage reaches sufficient levels, animals coming into contact with grounded devices may receive a mild electrical shock that can cause a behavioural response. At voltage levels that are just perceptible to the animal, behaviours indicative of perception such as flinches may result, with little change in normal routines.

Studies by numerous independent research groups in several countries are in agreement that the most sensitive cows begin to react to 50 or 60 Hz electrical current of 2 mA applied from muzzle to hooves or from hoof to hoof. This corresponds to a contact voltage level of about 1 V. As the voltage and current is increased, a greater percentage of cows will react with behavioural responses becoming more pronounced. Numerous studies have documented avoidance behaviours at levels above the first reaction threshold.

The median avoidance threshold for 50 or 60 Hz current flowing through a cow is about 8 mA. Even when the threshold is exceeded not all cows would be expected to show a behavioural response but as the voltage increases, signs in a herd would be expected to be more widespread and uniform. The scientific evidence strongly suggests there is no relationship between behavioural responses to stray voltage and physiological or hormonal responses.

There is no apparent relationship among behavioural modifications, milk production, and animal health. The only studies that have documented adverse effects of voltage and current on cows had both sufficient current applied to cause aversion and forced exposures. It is typical for voltage levels to vary considerably at different locations on a farm. Decreased water and feed intake or undesired behaviours will result only if current levels are sufficient to produce aversion at locations that are critical to daily animal activity.

These locations include feeders, waterers, and milking areas. Controlled research has shown that if an aversive voltage was administered to a water bowl once per second, water intake was reduced. However, when the same voltage was applied once every 10 min and once per day, no reduction in water intake was observed. If an aversive current occurs only a few times per day, it is not likely to have an adverse effect on cow behaviour.

The more often an aversive voltage occurs in areas critical to cows' normal feeding, drinking, or resting, the more likely it is to affect cows. No one sign is pathognomonic; a variety of signs has been reported in cows exposed to different levels of voltage. Documented signs are behavioural changes and decreased drinks of water per day and length of time per drink. The amount of water consumed may not be affected even when behavioural modification occurs.

Intermittent periods of poor performance, poor milk letdown, and incomplete or uneven milk-out, abnormal behaviour during milking, increased milking time, refusal of feed or water, increased SCC in milk, and increased mastitis are signs often attributed by farmers to stray voltage; however, none of these signs were evident in numerous controlled studies. These signs are often caused by other factors such as abusive cow handling, faulty milking machine, poor milking techniques and hygiene, and nutritional deficiencies.

Therefore, animal behaviour or other symptoms cannot be used to diagnose stray voltage problems. The only way to determine if stray voltage is a potential cause of abnormal behaviours or poor performance is to perform electrical testing. A thorough investigation of the entire production unit should be conducted to determine other sources of problems. Electrical systems should comply with wiring codes and standards at all times to protect both animals and people.

Whenever suggestive signs cannot be attributed to other causes, measurements should be taken to determine if a voltage potential exists, and the results recorded for future comparisons. A diagnostic confirmation of stray voltage must include a competent electrical measurement indicating at least 2 to 4 V between 2 points that a cow might contact, with some cows should exhibiting avoidance behaviours at this location. Voltage levels may need to be monitored at different times of the day and on different days because the threshold level may be exceeded intermittently.

All voltage readings should be made with a 500 to 1,000 Ω resistor across the 2 measuring leads to the cow contact points in addition to open circuit measurements. Readings without the use of a shunt resistor are meaningless. Although the resistance of cow and human tissues is similar, the contact resistance is generally lower for cows than for humans, particularly if cows are in a wet environment. The resistance of a cow's body plus the contact resistance with the floor is commonly estimated as 500 Ω. This is a reasonable value for a cow standing on a wet floor.

Cows standing on a dry surface will typically produce 1,000 Ω resistance or higher. Cows standing or lying on dry bedding will have a resistance many times higher than this. The resistance of a human can be as low as 1,000 Ω for wet hand-foot contact to >10,000 Ω for dry handfoot contact. The contact voltage to produce sensation can therefore be higher for humans than for cows, depending on the conditions of the contact points. If more than 1 V is detected at the cow contact points, it is advisable to have a qualified electrician or the local power supplier evaluate the situation.

CATTLE AND BUFFALO DEVELOPMENT

The broad frame-work of the cattle and buffalo breeding policy being followed since the mid-sixties envisaged selective breeding of indigenous breeds in their breeding tracts and use of such improved breeds for upgrading of the non-descript stock. While the States accepted the framework, appropriate implementation through field level programmes could not be done. Lack of interest in promoting Breed Organisation/Societies and related farmers' bodies contributed to the gradual deterioration of indigenous breeds.

Government intervention for breed improvement is not available to majority of owners of indigenous breeds of cattle. Eventually, the availability of good quality bulls needed for natural mating in breeding tracts became scarce, leading to further deterioration of indigenous breeds in these tracts. Production of quality indigenous bulls has been a long-neglected area and would require a major thrust in order to harvest the best male germplasm available in the country. The present production capacity of frozen semen doses is about 30 million against the estimated requirement of 65 million doses annually.

Except for a few pockets in important breeding tracts and in sperm stations, indigenous bulls of unknown pedigree and with poor quality semen are generally used. Crossbreeding, which was to be taken up in a restricted manner and in areas of low producing cattle, has now spread indiscriminately all over the country. Continuous emphasis on cross breeding with exotic breeds even in the tracts of indigenous breeds led to the nearextinction of some of the known breeds. Further, the indiscriminate use of contaminated semen or infected bulls results in the spread of sexually transmitted diseases like Infectious Bovine Rhinotracheitis (IBR) at an alarming rate.

MILK PRODUCTION

Milk production in India remained more or less stagnant from 1950 to 1970. Thereafter, it increased rapidly, reaching 84.6 million tonnes (mt) in 2001-02 (anticipated). But the Ninth Plan target of milk production (96.49 mt) was not achieved. The per capita availability of milk increased from 112 gm per day in 1973-74 to about 226 gm per day in 2001-02. However, it is still below the world average of 285 gm per day. Investment in the dairy sector in the Ninth Plan decreased significantly compared to the Eighth Plan.

Out of 168 Milk Unions, 58 Milk Unions (34.5 per cent) were running in loss as of March 2000. So far, the Government policy in the dairy sector has been to give preference to the establishment of milk processing plants linking rural milk producers to urban consumers through a network of cooperatives. Restrictions on establishing new milk processing capacity under Milk and Milk Products Order (MMPO) has now been removed. No policy measures have been undertaken so far to give a fillip to the unorganised sector involved in the production of Indian dairy products (like ghee, paneer, chhena, khoa etc.), which have tremendous potential in the export market in Asian and African countries.

EGG PRODUCTION

The Indian poultry industry has come a long way–from a backyard activity to an organised, scientific and vibrant industry. It is estimated that the egg production in the country is about 33. 6 billion numbers against the Ninth Plan target of 35 billion numbers. The most notable growth among the livestock products has been recorded in eggs and poultry meat. Since 1970-71, their output has grown at 5.87 per cent per annum.

The significant achievement in poultry development has come from the initiatives taken up by the private sector for commercial pure-line breeding. However, despite the huge investment made, mostly by the private sector, the poultry-processing sector is incurring losses. The status of the poultry sector as to; whether it falls under agriculture or industry, is somewhat ambiguous and, therefore, it has remained deprived of various benefits available to these sectors. Poultry farming should be declared as an agricultural activity.

The poultry production model in vogue (high input-high output using commercially developed strain of birds) has been primarily responsible for the rapid growth in production of eggs and broiler meat in the country, but it is successful mainly in large-scale units (more than 1,000 units of birds). Due to high feed cost, non-availability of credit and marketing support, most of the small farmers have become contract farmers and are exploited by middlemen.

Government intervention, by way of various support mechanisms, is now needed for the promotion of poultry in rural areas. Indigenous poultry breeds, including the improved strains that can survive with low quality raw feed and better resistance against diseases, can be reared under free range conditions by rural unemployed youth and women for some additional income and employment.

MEAT PRODUCTION

In India, meat production is largely a byproduct system of livestock production utilising spent animals at the end of their productive life. Meat production was estimated at 4.6 mt in 1998. Projects sanctioned during the Seventh and Eighth Plans for improvement/modernisation of abattoirs and carcass utilisation centres for fallen animals are still to be completed.

GOAT DEVELOPMENT

Despite the least attention from the planners, goat population in India has increased at the fastest rate among all major livestock species during last two decades. However, instead of increasing the goat population, emphasis should be laid on productivity per animal, organised marketing and prevention of emergence of new diseases like Peste des petits ruminants (PPR) which has led to higher mortality and abortion in goats. The goat improvement programme is to be given a push through extending credit to the poor landless farmers.

SHEEP PRODUCTION

During the last four decades, there has not been much increase in the sheep population. Production of wool has increased from 43.3 million kg in 1996-97 to 49.0 million kg (anticipated) in 2001-02. The Ninth Plan target of wood production (54.0 million kg) was not achieved. The fine wool production in the country is around 4 million kg against the demand of around 35 to 40 million kg. Indian wool is primarily used for the production of carpet, drugget, wall hangings etc. To enhance the quality and quantity of carpet wool, shepherds need incentives like credit, health coverage, breed improvement programmes and timely disposal of wool and surplus animals at a reasonable price.

PIG DEVELOPMENT

Pig husbandry is the most important activity in the animal husbandry sector in the northeastern region inhabited by tribal people. The region also has a substantial pig population, which constitutes around 25 per cent of the country's pig population. The bulk of the population is, however, of the indigenous type whose growth and productivity is very low. The major difficulty in pig development is the acute shortage of breeding males.

ANIMAL HEALTH

Since the Second Plan, efforts have been made to control diseases namely, Rinderpest, Foot and Mouth Disease, Haemorrhagic Septicemia, Black Quarter and Anthrax. Although Rinderpest has been eradicated from the country, the prevalence of the other diseases continues to be one of the major problems in the animal production programme.

Some of the emerging diseases like PPR, Bluetongue, Sheep Pox and Goat Pox, Classical Swine Fever, Contagious Bovine Pleuropneumonia, New Castle Disease (Ranikhet Disease) are causing substantial economic losses. The programme for creation of diseasefree zones was sanctioned in the Ninth Plan but was not implemented. The Department of Animal Husbandry and Dairying is also not well equipped with the necessary infrastructure and qualified technical manpower to execute the programmes and perform its mandatory duties and responsibilities like disease diagnosis and accreditation as per the international standards, development of an effective surveillance and monitoring system for diseases, mass immunisation against the most prevalent diseases etc.

Dovetailing the Animal Research Institutes of the Indian Council of Agricultural Research (ICAR) with the Department would not only improve its efficiency but also provide it with an effective delivery machinery to carry out its regulatory and certification authority functions, including the conservation of endangered breeds of livestock. The suggestion for the establishment of an Indian Council for Veterinary and Fisheries (ICVFR) by carving out the animal science and fishery institutes from ICAR has not yet materialised.

ANIMAL STATISTICS

The Livestock Census Scheme suffers from quantitative as well as qualitative problems. The present arrangements for conducting the Livestock Census in the States and Union Territories are not satisfactory in relation to timely collection of data and reporting. The Integrated Sample Survey Scheme for estimation of production of major livestock products also needs improvement.

CONSERVATION

The last few decades have witnessed serious erosion, and even extinction, of some indigenous animal breeds in the country. Many existing breeds are facing varying degrees of threat, endangerment and are heading towards eventual decimation. In all States, crossbreeding of cattle is now occupying a dominant position in the production programme and, in this process, the native cattle breeds, which are well adapted, have suffered wilful neglect resulting in their progressive elimination from the production system.

India is bestowed with rich domestic animal biodiversity, having 30 breeds of cattle, 12 breeds of buffalo, 20 breeds of goats, 40 breeds of sheep, eight breeds of camel, six breeds of horses, three breeds of pig and 18 of poultry. Besides, there are other species like equine, mithun, yak, turkey, ducks, etc. Indigenous breeds/types are rich in variability and are endowed with many positive traits like superior disease resistance, better tolerance to high heat and humidity and other characteristics suitable to particular agro-climatic environments.

It has also been noted that indigenous breeds are more efficient in feed conversion particularly the crop residues and naturally available low quality roughages.

Indigenous breeds at risk are:

- Cattle:
  - Red Sindhi,
  - Sahiwal,
  - Tharparkar,
  - Punganur and
  - Vechur.
- Buffaloes:
  - Nili-Ravi,
  - Bhadawari and
  - Toda.
- Sheep:
  - Nilgiri,
  - Muzaffarnagri,
  - Malpura,
  - Chokla,
  - Jaisalmeri,
  - Munjal,
  - Changthangi,
  - Tibetan,
  - Bonpala from Sikkim and
  - Garrole sheep
- Goat:
  - Beetal,
  - Jamunapari,

  - Chegu,
  - Changthangi,
  - Surti and
  - Jakhrana.
- Camel:
  - Bacterian,
  - Jaisalmeri and Sindhi.
  - Yak Mithun
- Poultry:
  - All the 18 indigenous breeds of poultry are facing extinction.
  - The three important breeds are Aseel, Kadaknath and Naked Neck,

It has been globally recognised that conservation and improvement of native animal genetic resources are essential for sustainable development in agriculture and animal husbandry.

The conservation and improvement programme should be decentralised and each State/adjoining States where a breed exists should take necessary steps with the active involvement of institutions, Gaushalas, Non-government Organisations (NGOs) and Breed Societies. The efforts should, however, be effectively coordinated centrally.

Given the severity of the resource constraint, all the Central sector and Centrally sponsored schemes were subjected to zero-based budgeting during the Ninth Plan.

The objective was to retain only those schemes that are demonstrably efficient and essential. The schemes that are similar in nature would be converged to eliminate duplication and resource flow would be linked to performance. Out of 41 schemes, 23 schemes were weeded out, one scheme was transferred and six schemes were merged.

Table. Average Annual Growth Rate of Milk and Egg Production 1950-51 to 2000-01

Year	Milk (%)	Eggs (%)
1950-51 to 1960-61	1.64	4.63
1960-61 to 1973-74	1.15	7.91
1973-74 to 1980-81	4.51	3.79
1980-81 to 1990-91	5.68	7.80
1990-91 to 2000-01	4.21	4.46

APPLICATION OF SENSORY ANALYSIS TO DAIRY PRODUCTS

Following are specific examples of how QDA and/or other types of sensory analysis techniques have been applied to dairy research studies funded by Dairy Management Inc. ™ (DMI).

OPTIMIZATION OF CHEDDAR CHEESE TASTE IN MODEL CHEESE SYSTEMS

Cheddar cheese, the most popular natural cheese in the United States, has a very complex flavour system. While much information has accumulated during the past century, the industry is still seeking to fully understand Cheddar cheese flavour and has not been able to replicate it in model systems. The non-volatile sensory attributes of Cheddar cheese are important for providing the character of Cheddar cheese. While much work has been published on volatile components of Cheddar, far less is known on how non-volatiles impact Cheddar flavour. Yang, a Kraft Foods researcher, and Vickers, at the Minnesota-South Dakota Dairy Foods Research Center, used sensory analysis to better understand the importance of non-volatile compounds to Cheddar flavour.

A descriptive panel was trained to evaluate real and model cheese for a variety of taste attributes and for Cheddarlike taste. Sodium chloride, lactic acid, citric acid and monosodium glutamate were added to the model systems using mixture designs and response surface methodology to determine optimum levels of these components.

The three model systems investigated were:

1. A dairy model system (containing milk isolate, anhydrous milk fat, water, annatto colour and chymosin);
2. A non-dairy model system (containing gelatin, gum acacia, modified starch, sunflower oil, water and annato colour); and
3. A mozzarella base.

While the mozzarella base did present tastes, it was used because the other two model systems were too unlike Cheddar cheese (or any cheese) texture. Less sodium chloride and fewer acids were required to simulate the taste of mild Cheddar compared with aged Cheddar. None of the model systems mimicked the texture of real Cheddar. The researchers were able to match approximately, but not exactly, the taste of aged Cheddar using a mozzarella base.

Panelists generally rated the optimized taste in the dairy model system as more Cheddar-like than the optimized tastes in the non-dairy model. Two methods were used to measure how close a sample was to the Cheddar concept. One was by measuring the similarity of the sample to either mild Cheddar cheese taste or aged Cheddar cheese taste on an unstructured scale. The left end of the line was marked with "not at all like Cheddar taste" and the right end was marked with "exactly like mild Cheddar taste" or "exactly like aged Cheddar taste." The other method was by concept matching using an R-index methodology. For the mild group, the panel evaluated whether the samples were "MC" (mild Cheddar taste and sure), "MC?" (mild Cheddar taste but not sure), "N?" (no Cheddar taste but not sure) or "N" (no Cheddar taste and sure).

For the aged group, "MC" and "MC?" were changed to "AC" (aged Cheddar taste and sure) and "AC?" (aged Cheddar taste but not sure).

A model system for studying Cheddar taste should be as bland-tasting as possible, and also have a texture and composition similar to that of real Cheddar cheese. The characteristic flavour of a food depends not only on the flavour compounds present and their levels but also the rate and extent to which they are released in real time, which in turn are affected by the amounts of proteins, fat and other matrix components of the sample.

By using a trained descriptive analysis panel, Yang and Vickers were able to evaluate the flavour impact of several non-volatile Cheddar cheese components (*i.e.*, salt, lactic acid, citric acid and monosodium glutamate) in model systems that attempted to mimic real Cheddar cheese. They achieved the most Cheddar-like taste with the mozzarella cheese base, and panelists found the optimal concentration of salts and acids in the model to be nearly indistinguishable from real Cheddar cheese.

CHEDDAR CHEESE AND POWDERED MILK LEXICONS

M. A. Drake, at the Southeast Dairy Foods Research Center, developed and validated a descriptive language for Cheddar cheese flavour. For the project, 240 representative cheese samples were collected. Fifteen individuals from industry, academia and government participated in roundtable discussions to generate descriptive flavour terms. A highly trained descriptive panel (n=11) refined the terms and identified references.

Identification of chemical references was conducted with the assistance of K. Cadwallader at the University of Illinois. Instrumental analyses (gas chomatography/mass spectrometry, or GC/MS) were conducted to identify many flavour compounds that were responsible for specific flavours and off-flavours in Cheddar cheese. Twenty-four Cheddar cheeses were then presented to the panel to validate the proposed lexicon.

The panel differentiated the 24 Cheddar cheeses as determined by univariate and multivariate analysis of variance. Twenty-seven terms were identified to describe Cheddar flavour. Seventeen descriptive terms were observed in most Cheddar cheeses. Drake's standard sensory language for Cheddar cheese today is facilitating training and communication among different research groups.

The Cheddar cheese lexicon is helping cheese-makers and cheese users accurately and consistently characterize the flavour of their cheese products and improve quality issues by measuring and controlling the presence of compounds that have been associated with flavour defects.

Following development of the Cheddar cheese lexicon, Drake developed a similar language to help characterize another food industry staple: dried dairy ingredients, including whey proteins and non-fat dry milk. Global production of non-fat dry milk tops 3.3 million tons and whey protein demand still outstrips production, which increases annually.

A sensory lexicon describing the flavour of these ingredients helps dairy processors maximize the quality of these ingredients and allows food technologists to identify the exact attributes or flavour notes these ingredients contribute to formulations. Drake said she was surprised by the number of descriptive terms that the panel uncovered for application to the dried dairy ingredients lexicon. The panel discovered 21 flavour terms that could be applied to milk powders. Examples included cooked/milky flavour, cake mix or vanillin, sweet and sour, earth and cereal.

Each of these flavours was linked to a key aroma compound, many of which were identified by Drake and Cadwallader with GC/MS. For example, lactones tend to lend a sweet, coconut like flavour, while various free fatty acids can simulate a waxy flavour. Many different factors contribute to flavour variability. The source of the powder, processing/packaging methods and materials, as well as storage time and conditions, are just a few.

DAIRY PRODUCTS

Dairy products are generally defined as foods produced from cow's or domestic buffalo's milk. They are usually high-energy-yielding food products. A production plant for such processing is called a dairy or a dairy factory. Raw milk for processing mostly comes from cows and to a lesser amount from domestic buffalos, but occasionally from other mammals such as goats, sheep, yaks, or horses. Dairy products are commonly found in European, Middle Eastern and Indian cuisine, whereas they are almost unknown in East Asian cuisine.

TYPES OF DAIRY PRODUCTS

- Milk after optional homogenization, pasteurization, in several grades after standardization of the fat level, and possible addition of bacteria Streptococcus lactis and Leuconostoc citrovorum
  - Crème fraîche, slightly fermented cream
  - a. Smetana, Central and Eastern European variety of sour cream
  - b. Clotted cream, thick spoonable cream made by heating
  - Cultured buttermilk, fermented concentrated (water removed) milk using the same bacteria as sour cream
  - Kefir, fermented milk resembling buttermilk but based on different yeast and bacteria culture
  - Kumis/Airag, slightly fermented mares' milk popular in Central Asia
  - Milk powder (or powdered milk), produced by removing the water from milk

a. Whole milk products
b. Buttermilk products
c. Skim milk
d. Whey products
e. Ice Cream
f. High milk-fat and nutritional products (for infant formulas)
g. Cultured and confectionery products

- Condensed milk, milk which has been concentrated by evaporation, often with sugar added for longer life in an opened can
- Evaporated milk, (less concentrated than condensed) milk without added sugar
- Ricotta cheese, milk heated and reduced in volume, known in Indian cuisine as Khoa
- Infant formula, dried milk powder with specific additives for feeding human infants
- Baked milk, a variety of boiled milk that has been particularly popular in Russia

• Butter, mostly milk fat, produced by churning cream
  - Buttermilk, the liquid left over after producing butter from cream, often dried as livestock food
  - Ghee, clarified butter, by gentle heating of butter and removal of the solid matter
  - Smen, a fermented clarified butter used in Moroccan cooking.
  - Anhydrous milkfat
• Cheese, produced by coagulating milk, separating from whey and letting it ripen, generally with bacteria and sometimes also with certain molds
  - Curds, the soft curdled part of milk (or skim milk) used to make cheese (or casein)
  - Whey, the liquid drained from curds and used for further processing or as a livestock food
  - Cottage cheese
  - Quark
  - Cream cheese, produced by the addition of cream to milk and then curdled to form a rich curd or cheese made from skim milk with cream added to the curd
  - Fromage frais
• Casein
  - Caseinates
  - Milk protein concentrates and isonates
  - Whey protein concentrates and isonates

  - Hydrolysates
  - Mineral concentrates
- Yogurt, milk fermented by Streptococcus salivarius ssp. thermophilus and Lactobacillus delbrueckii ssp. bulgaricus sometimes with additional bacteria, such as Lactobacillus acidophilus
  - Ayran
  - Lassi
- Clabber (food), milk naturally fermented to a yogurt-like state
- Gelato, slowly frozen milk and water, lesser fat than ice cream
- Ice cream, slowly frozen cream and emulsifying additives
  - Ice milk
  - Frozen custard
  - Frozen yogurt, yogurt with emulsifiers that is frozen
- Other
  - Viili
  - Kajmak
  - Filmjölk
  - Piimä
  - Vla
  - Dulce de leche

HEALTH RISKS OF CONSUMING DAIRY PRODUCTS

Most dairy products contain large amounts of saturated fat. Some dairy products may cause health issues for individuals who have a lactose intolerance and milk allergies. Some dairy products such as blue cheese may become contaminated with the fungus Aspergillus fumigatus during ripening, which can trigger asthma and other respiratory problems in susceptible individuals.

Vegans and some vegetarians avoid dairy products due to a variety of ethical, dietary, environmental, political, and religious concerns.

PASTURE USE IN ORGANIC MILK PRODUCTION

The most common technology used on conventional dairy operations confines milk cows in large barns and limits access to pasture. This means that forage from grazing comprises very little of the forage fed to dairy cows on most conventional dairies. Access to pasture is a requirement for the organic certification of dairy operations. Forage from grazing is also an important element of the feeding programme on many organic dairies.

Milk producers may find it easier and less expensive to maintain organic pastures for grazing dairy cows than to either purchase or produce and harvest organic crops and forage for dairy feed. Also, grazing systems may be less stressful for dairy cattle and contribute to lower veterinary expenses. Pasture-based production may also have environmental benefits, such as improved soil quality and reduced soil erosion. Pasture use for dairy feed on organic dairies, relative to conventional dairies.

More than 60 per cent of organic milk producers reported that at least half of their total forage ration came from pasture during the grazing months (an average of 6.5 months per year), compared with 18 per cent of conventional dairies. Nearly 90 per cent of organic dairies sourced at least 25 per cent of their total forage ration from pasture. In contrast, 70 per cent of conventional dairies reported that less than 25 per cent of their total forage ration fed to dairy cows came from pasture, while 40 per cent obtained none from pasture.

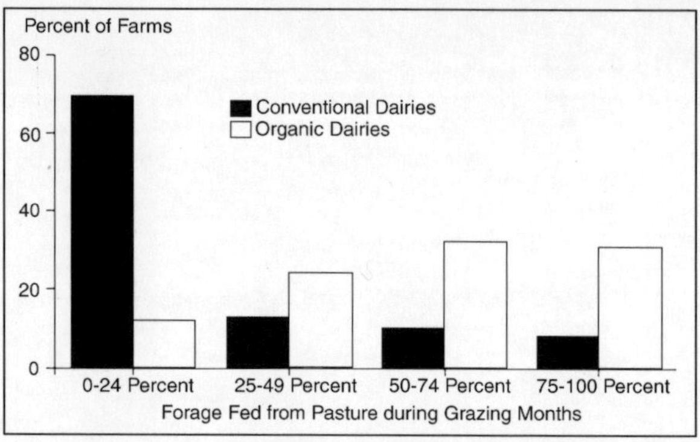

Fig. Producer Type, by Share of Dairies Forage Fed from Pasture

Notes:

More than 60 per cent of organic dairies obtained at least 50 per cent of the forage fed to dairy cows from pasture, while nearly 70 per cent of conventional dairies obtained less than 25 per cent

To evaluate the role of pasture grazing in organic milk production, surveyed producers were divided into groups that indicate the level of pasture grazing as a source of dairy cow forage during the grazing months:

- 0-24 per cent;
- 25-49 per cent;
- 50-74 per cent; and
- 75-100 per cent.

CHARACTERISTICS AND PRACTICES BY LEVEL OF PASTURE USE

Almost two-thirds of organic milk producers reported that at least 50 per cent of the forage fed to dairy cattle during the grazing months came from pasture, and a third reported that 75 per cent or more came from pasture. Only 12 per cent of organic milk producers reported that less than 25 per cent of forage fed came from pasture. Differences between organic dairies where pasture use was highest (75-100 per cent of forage fed) and lowest (0-24 per cent of forage fed) were significant and are emphasized in this section.

Farm characteristics among organic dairies in the middle groups (25-49 per cent and 50-74 per cent of forage fed from pasture) showed little difference. Operations that relied on pasture use the most were the smallest in size and produced the least milk per cow. Organic dairies with 75-100 per cent of forage fed from pasture had an average of 64 milk cows that averaged 11,289 pounds of milk per cow.

Table. Characteristics and Production Practices, by Level of Pasture use on Organic Dairy Operations, 2005

	Forage Fed from Pasture During Grazing Months			
Item	**0-24** %	**25-49** %	**50-74** %	**75-100** %
Per cent of farms	12	24	32	31
Per cent of milk cows	21	23	31	25
Per cent of milk production	25	23	31	21
Milk cows *(number per farm)*	135	79	78	64
Milk production *(pounds per cow)*	16,560	13,594	13,319	11,289
Region:		*Per cent of farms*		
Northeast (ME, NY, PA, VT)	21	42	43	55
Upper Midwest (MI, MN, WI)	58	44	40	35
Corn Belt (IL, IN, IA, MO, OH)	13	8	5	9
West (CA, ID, OR, WA)	8	6	12	2
Region:		*Per cent of cows*		
Northeast (ME, NY, PA, VT)	9	29	28	45
Upper Midwest (MI, MN, WI)	30	46	28	29
Corn Belt (IL, IN, IA, MO, OH)	10	9	4	8
West (CA, ID, OR, WA)	51	16	40	18
In dairy business:		*Per cent of farms*		
Less than 10 years	11	18	19	27
10 years or more	89	82	81	73
Pasture rotation:				
At least once a day	39	48	58	72
Every 2-7 days	30	33	38	26
Less than weekly	4	1	id	1
Never	24	18	3	1
Pasture not used for feed	4	0	0	0

Notes:

id=Insufficient data for disclosure.

In contrast, dairies with 0-24 per cent of forage from pasture averaged more than twice the number of milk cows (135) and more than 5,000 pounds more milk production per cow (16,560 pounds). Thus, operations that relied less on pasture use (12 per cent) had a disproportionately large share of milk cows (21 per cent) and milk production (25 per cent), while a third of operations that relied most on pasture use accounted for only 25 per cent of milk cows and 21 per cent of production.

Organic dairies with the highest pasture use for feed were located in the Northeast. More than half of these farms were in the Northeast and operated with 45 per cent of the organic milk cows in this group. Operations in the Upper Midwest accounted for the majority of farms using the least pasture (0-24 per cent) and 35 per cent of farms using the most (75-100 per cent). Organic dairies in the West accounted for only 8 per cent of the farms using the least pasture, but these large operations held 51 per cent of the milk cows in this group.

Large operations relied less on pasture as a forage source because of the significant land requirements necessary to supply pasture for large herds. However, some large operations in the West used pasture as an important dairy feed source. Among the farms with the highest pasture use, the West (at 2 per cent) included 18 per cent of the milk cows. Organic dairies that relied the most on pasture use were in business relatively less time than other organic dairies.

Twenty-seven per cent of dairies that relied on pasture for 75-100 per cent of their forage feed had been in business less than 10 years, compared with only 11 per cent that relied on pasture for 0-24 per cent of their forage feed. Also, farm operators who relied the most on pasture forage were younger than other farm operators.

Sixteen per cent of these farm operators were younger than 35 and only 4 per cent were older than 65, compared with 6 and 10 per cent, respectively, of the operators who used the least pasture. Younger farm operators on relatively new dairies were more likely to use pasture-based feeding for organic milk production.

Pasture as a source of forage on organic dairies was closely related to the intensity of pasture management. Rotating pastures is a management strategy that can increase the total volume and quality of dry matter produced on pastures by allowing cows to graze for specified periods before moving them to new pastures.

Seventy-two per cent of organic dairies that reported the highest use of pasture forage rotated pastures at least once a day, compared with less than 40 per cent of dairies that used pasture forage the least. Among dairies that used pasture the least, 24 per cent reported that pastures were never rotated and 4 per cent indicated that pasture was not a source of dairy feed.

PRODUCTION COSTS AND NET RETURNS BY LEVEL OF PASTURE USE

Substituting pasture for more expensive feed sources may appear to be an efficient way to lower dairy feed costs. Average feed costs per cwt of milk, however, were lowest on the organic operations using the least pasture for dairy forage (0-24 per cent), between $1 and $2 per cwt less than on the operations using more pasture.

Table. Production Costs and Net Returns on Organic Dairy Operations, by Level of Pasture use, 2005

Item	Forage Fed from Pasture During Grazing Months			
	0-24 %	25-49 %	50-74 %	75-100 %
	Dollars per cwt sold			
Feed:	11.61	13.14	13.42	12.90
Purchased feed	7.43	5.15	7.19	6.09
Homegrown harvested feed	3.89	7.51	5.41	5.69
Grazed feed	0.29	0.48	0.82	1.13
Hired labour	2.44	1.85	2.17	2.27
Capital costs	4.11	5.36	4.48	6.70
Opportunity cost of unpaid labour	2.82	6.47	6.24	8.36
Cost summary				
Operating costs	17.74	18.55	18.99	19.44
Operating and capital costs	21.85	23.91	23.47	26.14
Total economic costs	25.33	31.53	30.64	35.99
Gross value of production	24.10	24.49	23.59	25.63
Returns above total economic costs	-1.22	-7.04	-7.05	-10.36

Notes:

Measuring Milk Production Costs. The gross value of production includes milk sales, cull and breeding animal sales, revenue from leasing dairy animals or space to other operations, co-op patronage dividends, dairy assessment rebates or refunds, and the value of manure. The value of milk sales was determined from prices reported by the survey respondents.

This relationship changes little even if charges for pasture resources are excluded from the feed costs. Average feed costs per cow declined as pasture use for dairy forage increased. Total feed costs were $1,902 per cow on organic dairies that relied on pasture for 0-24 per cent of forage fed, compared with only $1,409 on organic dairies that relied on pasture for 75-100 per cent of forage fed, a savings of about 25 per cent.

Breaking down the distribution of milk production per cow for each group provides an indication of how milk yields declined as pasture was substituted for higher energy feed in the dairy ration. Nearly half of organic dairies feeding 0-24 per cent of forage from pasture had annual milk yields at 15,000 pounds per cow or more, compared with just 10 per cent of dairies feeding 75-100 per cent.

In contrast, more than 40 per cent of dairies using the most pasture had an annual milk yield of less than 10,000 pounds per cow, compared with 19 per cent of those using the least pasture. This suggests that improving pasture quality to achieve higher milk production can contribute to the success of organic dairies using pasture-based feeding. The relationship between pasture use and total production costs follows much the same pattern as seen for size and production costs on organic dairies. That is, the operations using less pasture were much larger and had lower average costs per cwt than the smaller operations using more pasture.

Capital costs were more than $2 per cwt higher and the charge for unpaid labour was more than $5 per cwt higher on the operations where pasture forage accounted for 75-100 per cent of forage fed than on those where pasture forage accounted for 0-24 per cent of forage fed.

Average returns net of total economic costs were negative for dairies in each of the pasture use groups, but were significantly higher among the dairies using the least pasture. This disparity suggests that organic operations using conventional dairy feeding methods, such as confining cows and feeding higher energy feed, were more likely to generate higher returns to capital and labour resources than those relying more on pasture-based feeding.

Thus, the technologies and production methods used on organic dairies could become more like those used on conventional dairies. Changing technologies and production methods may depend on the extent to which pasture requirements for organic dairies must include pasture as part of the dairy feeding programme.

LABOUR USE BY LEVEL OF PASTURE USE

The biggest difference in organic dairy costs between those using the most and the least pasture for dairy forage was the unpaid labour charge. The labour charge on organic dairies feeding 75-100 per cent of forage from pasture was more than $8 per cwt, compared with less than $3 per cwt on dairies feeding 0-24 per cent of forage from pasture. The labour charge contributed about half the difference in total costs that were more than $10 per cwt higher for dairies using the most pasture. Table includes labour use per cow and per cwt for each level of pasture feeding.

Unpaid labour use per cwt on organic dairies feeding 75-100 per cent of forage from pasture was about three times that for those feeding 0-24 per cent of forage from pasture. Part of this difference came from fixed amounts of labour being spread over more output on the larger operations that used the least amount of pasture. Some of the labour difference, however, may be due to the pasture feeding system's being more labour intensive than handling dairy cows in confinement.

Among farms with fewer than 100 cows, the unpaid labour use on dairies feeding the least forage from pasture was still about a third less than on those feeding the most forage from pasture (0.45 hour versus 0.65 hour). In addition, total labour use among these small dairies increased steadily as more pasture was used for dairy feed. This increase suggests that pasture-based feeding systems are inherently more labour intensive than the conventional practice of handling dairy cows in confinement, possibly due to the labour required for moving cows to, from, and between pastures.

Table. Organic Dairy Labour Use, by Level of Pasture Use, 2005

	Forage Fed from Pasture During Grazing Months			
Item	**0-24** %	**25-49** %	**50-74** %	**75-100** %
Per cow				
Hired labour hours	30	20	22	17
Unpaid labour hours	27	52	48	56
Total labour hours	57	72	70	73
Per cwt				
Hired labour hours	0.18	0.15	0.17	0.15
Unpaid labour hours	0.16	0.39	0.36	0.49
Total labour hours	0.34	0.53	0.52	0.65
Per cwt—dairies with fewer than 100 cows				
Hired labour hours	0.15	0.18	0.17	0.16
Unpaid labour hours	0.45	0.52	0.59	0.65
Total labour hours	0.60	0.70	0.76	0.81

Note:

Totals may not equal the sum of items due to rounding

ECONOMIES OF SIZE IN ORGANIC MILK PRODUCTION

The emergence of large conventional dairy farms and the continued shift of production towards such farms suggest that economies of size play a major role in the structure of dairy farming. Economies of size and the role it takes in the structural changes observed in conventional milk production were documented by MacDonald *et al.* This section reports on the relationship between operation size and production costs in organic milk production, including what size and costs might mean for the structure of the industry. To evaluate the relationship between organic milk production costs and operation size, surveyed producers were divided into groups by size, and differences in farm characteristics and practices, production costs and net returns, and labour use were compared among the groups.

The size groups were defined by the largest number of cows milked on the operation during 2005:

- Fewer than 50 cows;
- 50-99 cows;
- 100-199 cows; and
- 200 cows or more.

CHARACTERISTICS AND PRACTICES BY SIZE

Data for 2005 indicated that about 45 per cent of organic dairies milked fewer than 50 cows, while 87 per cent had fewer than 100 cows. These small operations accounted for a disproportionately small share of organic milk cows and production. In contrast, about 13 per cent of organic dairies had 100 cows or more, but they accounted for 44 per cent of milk cows and nearly half of production.

The largest organic dairies—those with 200 cows or more—had an average herd size of almost 500 cows per farm and included nearly a third of organic milk cows. These large dairies had a higher average milk yield per cow than smaller dairies and produced 37 per cent of organic milk.

The most striking difference among organic dairies in each size group was their location. Northeast and Upper Midwest farms accounted for 96 per cent of dairies with fewer than 50 cows. Farms in the West accounted for 80 per cent of dairies with 200 cows or more. Within each region, 97 per cent of organic dairies in the Northeast, 88 per cent of organic dairies in the Upper Midwest, and 85 per cent of organic dairies in the Corn Belt had fewer than 100 cows. In contrast, 83 per cent of organic dairies in the West had 100 cows or more and 63 per cent had 200 cows or more.

The use of several milk production technologies and practices tended to increase with size of organic operation, but most of the differences were between the smallest and other organic farms. For example, use of a milking parlor increased from 22 per cent for the smallest dairies to nearly all dairies with 100 cows or more. Thirty-eight per cent of the smallest farms and over 50 per cent of other farms participated in the Dairy Herd Improvement programme.

Table. Dairy Farm Characteristics and Production Practices, by Size of Organic Operation, 2005

	Size of Organic Dairy			
Item	**Fewer than 50 Cows**	**50-99 Cows**	**100-199 Cows**	**200 Cows or More**
Per cent of farms	45	42	8	5
Per cent of milk cows	21	35	13	31
Per cent of milk production	18	33	12	37
Milk cows (number per farm)	37	68	132	490
Milk production (pounds per cow)	11,884	12,796	12,008	16,133
Per cent of farms				
Region:				
Northeast (ME, NY, PA, VT)	51	46	17	0
Upper Midwest (MI, MN, WI)	45	39	52	20
Corn Belt (IL, IN, IA, MO, OH)	4	12	15	0
West (CA, ID, OR, WA)	0	3	16	80
Production practices:				
DHI programme participation[1]	38	51	53	60
Pasture-based feeding[2]	70	60	49	54
Artificial insemination	70	76	71	87

Embryo transplants or sexed semen	3	3	6	2
Controlled breeding/calving season	38	35	36	20
Regular veterinary services	28	41	72	55
Nutritionist services	30	53	66	70
Computerized milking system	1	4	4	7
Computerized feeding system	id	3	5	31
Milking parlor	22	45	88	100
Kept individual cow records	59	64	62	70
Johne's disease programme participation	18	28	44	45
Onfarm computer records	13	25	34	41
Dairy information from Internet	30	46	56	64
Forward-purchased inputs	4	10	22	16
Negotiate input price discounts	18	21	33	36

Notes:

id=Insufficient data for disclosure.

[1]Dairy Herd Improvement.

[2]Pasture-based feeding is defined as providing at least half of the forage fed to milk cows during the grazing months from pasture. Organic dairies reported an average grazing period for milk cows of 6. 5 months in 2005.

Use of regular veterinary and nutritionist services was also much lower on the smallest farms. Likewise, the use of business management tools, such as computer records, the Internet, and forward-purchasing inputs, was less. Pasture-based feeding was used by 70 per cent of the smallest dairies and was also used by more than half of the largest dairies.

PRODUCTION COSTS AND NET RETURNS BY SIZE

Average feed costs differed little among the dairies in each size group, but average operating costs for organic dairies increased from the smallest to largest size group due to the additional costs of hired labour on the largest farms. Total operating costs were about $2 per cwt higher on the largest organic dairies than on the smallest. Once capital costs were added, cost differences among the size groups were negligible, as lower capital costs on the larger operations offset the additional hired labour costs.

Table. Production Costs and Returns, by Size of Organic Dairy Operation, 2005

	Size of Organic Dairy			
Item	**Fewer than 50 Cows**	**50-99 Cows**	**100-199 Cows**	**200 Cows or More**
Dollars per cwt sold				
Feed:	12.96	12.67	11.93	13.06
Purchased feed	5.16	4.94	5.29	8.99
Homegrown harvested feed	6.96	7.00	6.06	3.51
Grazed feed	0.83	0.73	0.58	0.56
Hired labour	0.79	1.75	2.25	3.21
Capital costs	6.02	6.20	5.04	3.58

Opportunity cost of unpaid labour	13.42	7.45	4.84	1.15
Cost summary:				
Operating costs	17.65	18.25	18.10	19.66
Operating and capital costs	23.67	24.44	23.13	23.25
Total economic costs	38.50	33.36	29.13	24.84
Gross value of production	25.57	24.88	23.46	23.59
Returns above total economic costs	-12.93	-8.48	-5.67	-1.26

Notes:

Costs are defined in Appendix A: Measuring Milk Production Costs. The gross value of production includes milk sales, cull and breeding animal sales, revenue from leasing dairy animals or space to other operations, co-op patronage dividends, dairy assessment rebates or refunds, and the value of manure. The value of milk sales was determined from prices reported by the survey respondents.

The smallest farms, however, used significantly more unpaid labour, mainly from farm operators and other family members, than did larger farms. After adding an opportunity cost for this unpaid labour, total economic costs for organic dairies declined with size and were nearly $14 per cwt less on the largest farms than on the smallest.

Average production costs for conventional milk production declined from about $30 per cwt on operations with fewer than 50 cows to about $18 per cwt on operations with 200-499 cows. Average costs dropped to less than $14 per cwt on conventional operations with more than 1,000 cows. Lower costs among the largest conventional producers suggest that similar economies of size may be available to organic producers if they increase the size of their operation above the industry norm.

However, the additional costs of complying with pasture requirements and securing organic inputs in large volume may limit the cost advantages for larger organic operations. Net returns to the organic dairy enterprise were computed as the difference between the gross value of production and total economic costs. Average net returns in 2005 were negative for all organic dairy size groups (from-$12.93 per cwt among dairies with fewer than 50 cows to-$1.26 per cwt among dairies with 200 cows or more).

Farmers are unlikely to start or expand a dairy operation if they are unable to recover their investment in capital and labour. Farmers who have already made substantial capital investments in dairy facilities and equipment, however, with little or no opportunity for use outside the dairy operation. These costs, therefore, have no bearing on the decision to operate in the short term, only becoming important when these capital assets must be replaced.

To better measure costs relevant to shortterm operating decisions, the share of farms with gross returns exceeding all costs except for capital costs was computed. About a fourth of organic dairies with fewer than 50 cows covered these costs, compared with almost threefourths of those with 200 cows or more.

Some small organic dairy farms appeared to be earning enough to keep operating, but on average, farms in the smaller size groups were not covering the opportunity costs of their investments in capital and the operator's time.

In contrast, larger organic dairies were much more likely to generate returns above capital and labour costs. These conditions suggest that organic milk production may migrate towards larger operations, as has conventional production. This production change will likely occur over an extended period as existing smaller operations use up their production facilities and equipment, ultimately facing decisions about replacing capital assets or exiting the industry.

Also, in the short term, some small operators may continue producing organic milk as a lifestyle choice despite returns that fail to cover the opportunity cost of their time.

LABOUR USE BY SIZE

Small organic dairies are at a cost disadvantage relative to larger dairies due to the opportunity cost assigned to the unpaid labour provided by operators, partners, and family members. Unpaid labour costs for organic dairies with fewer than 50 cows were $13.42 per cwt, compared with only $1.15 per cwt on dairies with 200 cows or more. If this opportunity cost is ignored, the net returns to organic milk production were more similar among the size groups. The largest dairies used substantially more hired labour than the smallest (11,530 hours versus 557 hours), but the hired labour hours per unit of production differed little among the size groups because of much higher production on the largest dairies.

The difference in unpaid labour use among the farm size groups was much less. Organic dairy operations with 200 cows or more used about 700 more hours of unpaid labour, about 20 per cent more than those with fewer than 50 cows. The largest dairies, however, produced about 1,700 per cent more milk as a result of having more cows and an average productivity per cow that was 36 per cent higher than the smallest dairies. Enhanced productivity for these larger dairies means that unpaid labour hours per unit of production fell dramatically across the size groups from 0.82 hour per cwt on the smallest organic dairies to only 0.05 hour on the largest.

Table. Labour Use, by Size of Organic Dairy Operation, 2005

	Size of Organic Dairy			
Item	**Fewer than 50 Cows**	**50-99 Cows**	**100-199 Cows**	**200 Cows or More**
Per farm				
Hired labour hours	557	1,648	3,051	11,530
Unpaid labour hours	3,612	3,759	4,265	4,322
Total labour hours	4,169	5,407	7,316	15,852

Milk production (cwt)	4,367	8,748	15,825	78,977
Per cwt				
Hired labour hours	0.13	0.19	0.19	0.15
Unpaid labour hours	0.82	0.43	0.27	0.05
Total labour hours	0.95	0.62	0.46	0.20

Unpaid labour is an opportunity cost defined as the time spent working on a dairy operation charged at a wage rate that represents what unpaid workers could earn in off-farm employment. Opportunity costs of labour may vary significantly among producers, and some producers may be willing to accept returns lower than they could earn in non-farm employment because of lifestyle preferences and costs of switching occupations, among other reasons. The incentive to earn a competitive return for the time spent on the dairy operation, however, may be a motivating force for increasingly larger organic dairies.

BEEF CATTLE

Beef cattle includes all animals of the genus *Bos* and their close relatives that are raised primarily for meat production. As ruminants, beef cattle are capable of utilizing a wide range of feedstuffs and consequently are maintained in an array of situations ranging from extensive grazing to confined feedlot pens and intensive laboratory environments. Regardless of the housing system, basic needs for food, water, shelter, and comfort should be met.

FACILITIES AND ENVIRONMENT

Thermal Indices

At the present time, the temperature-humidity index {THI; THI = 0.8 × ambient temperature + [(per cent relative humidity/100) × (ambient temperature–14.4)] + 46.4} has become the de facto standard for classifying thermal environments in many animal studies and selection of management practices during seasons other than winter. The THI, first proposed by Thom has been extensively applied for moderate to hot conditions, even with recognized limitations related to airspeed and radiation heat loads.

A THI between 70 and 74 is an indication to producers that the potential for heat stress in livestock exists. In particular, when THI values are above 70 by 0800h, it is recommended that managers of confined cattle that have high metabolic heat loads initiate or prepare to initiate heat-stress management strategies before cattle become exposed to the excessive heat load. A THI of 84 or above can cause death, especially in feedlot cattle that are within 45 d of slaughter and consuming high-energy finishing diets.

Modifications to the THI have been developed to overcome the shortcomings related to the lack of airflow and radiation heat load in the index. Eigenberg also developed similar adjustments based on predictions of respiration rates using ambient and dew point temperature, windspeed, and solar radiation. These models have merit in that the combined effects of multiple environmental factors can be taken into account when determining animal comfort. Gaughan developed a more extensive index as a guide to the management of feedlot cattle during hot weather. The heat load index (HLI) incorporates black globe temperature, relative humidity, and windspeed. A threshold above which cattle are less efficient at dissipating heat was developed for a reference animal.

The threshold for a full-blood Brahman steer is 96. Also, adjustments to the threshold are possible for use of shade, clean dry pens, cattle coat colour, and days on feed. The thresholds are lowered if cattle are sick or not acclimated to summer conditions. Very limited data exist for assessing environmental effects on reproduction.

However, Amundson found THI and daily minimum temperature to be equally good predictors of pregnancy rate at 42 d into the breeding season. However, the combination of wind speed and THI had the greatest correlation to pregnancy rate. Indices for cold stress are not as well defined as for heat stress. The wind chill index (WCI) has traditionally been used to derive an apparent temperature for humans. In 2001, the National Weather Service released a new WCI that may have merit for assessing effects of wind on domestic livestock.

Ideal Thermal Conditions

Under most environmental conditions, temperature represents a major portion of the driving force for heat exchange between the environment and an animal. However, moisture and heat content of the air, thermal radiation, and airflow also affect total heat exchange. Thus, a combination of environmental variables contributes to the conditions to which an animal responds. Under conditions in which relative measures and comparisons of the effect of different environmental variables could be determined, the apparent ambient temperature at which animals can cope has been defined with a reasonable degree of accuracy; however, variation does exist among animals.

Environmental conditions that provide maximum comfort and require little or no energy expenditure for maintenance depend on cattle age, metabolic size, and/or body mass and surface area. The TCZ generally ranges between 15 to 25°C for most cattle less than 1 mo old; between 5 and 20°C for a mature beef cow consuming a maintenance diet; and between–10 and 20°C for yearlings with ad libitum access to energy-dense feedlot diets.

Based on physiological responses and heat load thresholds, *Bos indicus* and some heat-tolerant *Bos taurus* cattle breeds have a TCZ at least 5°C greater than typical *Bos taurus* cattle. Encompassing the TCZ is the thermoneutral zone (TNZ). Within the TNZ, an animal can maintain homeostasis through normal physiological and metabolic processes, which may require minimal expenditure of energy when the animal is exposed to conditions outside the TCZ. The TNZ generally ranges between 10 and 30°C for most cattle less than 1 mo old; between–15 and 28°C for a mature beef cow consuming a maintenance diet; and between–35 and 25°C for yearlings with ad libitum access to energy-dense feedlot diets. Even though the upper end of the TNZ for most *Bos taurus* cattle is between 25 and 30°C, for high-producing cattle with high intakes of metabolizable energy the upper limit may be closer to 20°C on sunny days when little or no wind is present.

When given sufficient time, cattle acclimate and adapt to colder or hotter conditions. It should be noted that cattle that are adapted to–35°C may be uncomfortable at 10°C. Thus, the TCZ and TNZ serve only as guidelines to describe the limits within which cattle are comfortable and can adapt to, respectively.

Independent of these guidelines, performance standards that indicate a problem with the thermal environment include, in cold weather, shivering, huddling, and loss of body condition/weight; and in hot weather, panting, sweating, and a reduction in feed intake. Primary factors that affect thermal comfort include feed/energy intake and body condition/fat cover.

Range and Pasture Systems

Acceptable systems for grazing beef cattle on pasture and rangeland vary widely. Cow body condition is an excellent performance standard for monitoring the well-being and nutritional status of range cattle. Special consideration needs to be given to environmental factors that affect grazing beef cattle. In areas where heat stress is common, provision of shade to decrease the solar heat load is the most practical intervention in pasture and range systems.

The need for artificial shade should be assessed after careful consideration of the adequacy of naturally occurring sources. Heat stress is evidenced when respiration rates begin to increase. Prolonged increases in body temperatures will result in decreased feed intake, body condition, and weight. In areas where exposure to extreme cold is likely, provision of shelter for grazing beef cattle may be desirable.

Grazing beef cows decrease grazing time and forage intake as ambient temperature decreases below 0°C, although such changes are small in adapted beef cows. Cattle use windbreaks to decrease wind chill and prevent exposure to blowing snow, although it has not been clearly established that windbreaks improve animal performance.

Supplementary feed should be provided during periods of heavy snow cover that preclude grazing. An adequate supply of forage should be available to grazing cattle. Intake and performance may be decreased when the amount of standing forage is lacking but the appropriate quantity of forage dry matter per hectare varies with the pasture or range type and the stocking rate. Guidelines for acceptable amounts of standing forage per unit of body weight at given stocking rates are available but additional research is needed with a variety of pasture and range types. Grazing beef cattle should be provided with supplements for nutrients that are known to be deficient in pasture and range forage in particular localities. In almost all grazing environments, range cattle require free-choice access to supplemental salt as a source of supplemental sodium.

Typically, these salt-based, free-choice mineral supplements will also be fortified with trace minerals. Observation and monitoring of range cattle often occur less regularly than for other livestock. When supplemental feed is provided, cattle are usually observed at least 2 or 3 times weekly. Unsupplemented cattle on open range may be observed less frequently.

However, it is recommended that range cattle be observed at least once per week. In certain areas, grazing beef cattle may be affected by predators and poisonous plants. Careful attention should be given to such problems, and efforts should be made to decrease or eliminate these adverse conditions. Availability of fresh, unfrozen water is critical for grazing beef cattle, and distance to water should be given consideration in pasture and range systems.

If cattle are required to travel long distances to water in hot, dry climates, animal performance and utilization of pasture forage can be affected. Holechek recommended that distance to water be no greater than 1.6 km in rolling, hilly country and in undulating, sandy terrain. This recommendation was decreased to 0.8 km in rough country, increased to 2.4 km in smooth, sandy terrain, and increased to 3.2 km in areas with flat terrain. Thus, the distance to water for grazing cattle should not exceed 3.2 km, and every animal should have the opportunity to drink ad libitum at least once per day.

Feedlot and Housing Systems

Beef cattle used in research or teaching may be housed in intensive management systems, either indoors or in open lots, with or without shelter. Facilities for beef cattle should provide cattle with opportunities for behavioural thermoregulation. Management of dairy beef is similar to other cattle, although, some feeding, housing, and marketing regimens are unique to Holsteins. Proper airflow and ventilation are essential in intensive facilities.

In feedlots, cable or wire fencing has minimal effect on natural airflow in summer. However, high airflow rates are undesirable during periods of low temperature, and tree shelterbelts and other types of windbreak can decrease the rate of airflow past the cattle. An 80 per cent solid windbreak 3 m high decreases wind speed by half for about 45 m downwind and controls snow for about 8 m; a similar windbreak 4 m high decreases wind speed by half for about 65 m downwind and controls snow for about 10 m.

A windbreak is recommended in mounded, south-sloping feedlots in the northern United States to provide dry resting areas with low air velocities.

Caution should be exercised when placing cattle in sheltered areas in the summer because of the adverse effects of restricted airflow on cattle reared in hot environments. During potentially stressful heat episodes, panting scores (1 = elevated respiration rate, 2 = drool or saliva present on side of mouth, 3 = open mouth breathing observed, and 4 = tongue and neck extended with open mouth breathing) can be utilized as an excellent indicator of stress levels experienced.

When cattle are beginning to experience panting scores of 2 or greater some means of cooling may be needed. Cattle learn to take evasive action to alleviate heat stress and such competition for cooler areas in a pen or around the water trough increases even during cooler days in which heat alleviation methods are not utilized. When this occurs evidence of crowding is observed, which exacerbates the heat stress problems.

Wetting the ground or floor of holding facilities can be an effective method of cooling cattle managed in unshaded, outdoor units where surface vegetation is sparse or non-existent. Direct wetting of cattle during extreme heat is also an effective practice and is often used as an emergency measure. Benefits of sprinkling are enhanced if sprinkling is started in the morning, before cattle experience high heat loads. Generally, a daily application of 0.5 to 1.0 cm of water is sufficient to cool pen surfaces. However, applying 1.25 to 1.50 cm every other day is acceptable and will not sufficiently contribute to mud build-up in normally dry pens.

In areas with high evaporation rates additional water may be needed, which can serve to cool pen surfaces as well as eliminate potential dust problems. The size of the area needed to be sprinkled would be similar to the shade area recommendations. As a routine protective practice, wetting can be efficiently accomplished by utilizing a timer to provide 5 to 10 min of spray during each 20-to 30-min period. Fogger nozzles are often mistakenly recommended for wetting animals.

Fogger nozzles are less effective than sprinkler nozzles because of the barrier formed by the fine droplets. These droplets adhere to the outer hair coat of the animal, causing the heat for evaporation to come from the air rather than from the body. Mitlöhner *et al.* reported that misting cattle was not as effective as shade in decreasing heat stress, and in some cases, caused respiration rate to increase compared with non-misted cattle.

Shade for cattle can provide the margin of survival for animals that are unconditioned to a sudden heat wave with high solar radiant loads in central and southern regions of the United States. Mader found limited performance benefits of utilizing shade in the north-central region of the United States, in contrast to the findings of Mitlöhner where shade was effective in southern regions.

Also, use of shade in northern climates may be costly and logistically prohibitive because of snow load requirements, potential mud problems under shade and the low percentage of time that cattle may actually benefit from using the shade. However, benefits of using shade for maintaining animal comfort will almost always be found in any area or location in which abnormally hot or hot and humid conditions arise or persist, including northern climates, and when cattle have not had the opportunity to acclimate.

Mitlöhner *et al.* found excellent results when shades were provided for feedlot cattle reared in the south-central region of the United States, an area where more consistent benefits of shade would be expected to be realised. For optimum benefits shades should be 3.6 to 4.2 m high in areas with clear, sunny afternoons to permit maximum exposure to the relatively cool northern sky, which acts as a radiation sink.

In areas with cloudy afternoons, shades 2.1 to 2.7 m in height are more effective, as they limit the diffuse sky radiation received by animals beneath the shades. The amount of shade required for young cattle is 0.7 to 1.2 m^2 per animal, whereas larger cattle need 1.8 to 2.5 m^2 per animal. Shades are strongly recommended for sick cattle or for animals in hospital pens.

Cold housing can be provided for beef cattle. Open sides of any cattle building need to face away from prevailing winds. Such structures are ventilated by natural airflow, and the resultant winter temperatures are typically 2 to 5°C above outdoor conditions as a result of body heat. Totally enclosed housing requires ventilation to maintain the air temperature at acceptable levels and to minimize the accumulation in the air of water vapour, noxious gases, other odorous compounds, and dust. Ventilation systems may be either natural or mechanical.

Type of pen surface affects dustiness during hot dry weather and mud or manure build-up during wet periods. Good drainage of outside pens is imperative. Dirt pens should be regularly cleaned of animal waste residues and maintained to minimize accumulation of water. A hard surface apron in front of the feed bunks and around water troughs and shelters should be considered in dirt pens.

Mounds should be provided in dirt pens for cattle to lie on during inclement weather. Accumulation of mud in a pen or on the cattle can influence maintenance requirements and thermal balance. Properly designed pens with adequate slope are extremely important for minimizing mud and related health and behaviour problems. In areas where slope or drying conditions are limited, adding mounds is very useful for keeping cattle clean and dry. Under hot-humid conditions, mounds aid in preventing animal crowding and improve exposure to airflow for the animals that utilize them.

Additional information on feedlot/drylot pen design and layout has been published by Pohl and Henry. For hard-surfaced pens, materials should be durable, slip-resistant, and impervious to water and urine; easily cleaned; and resistant to chemicals and corrosion from animal feed and waste. Concrete floors should be scored or grooved during construction to improve animal footing. Properly designed slotted floors are self-cleaning. Fences, pen dividers, walls, gates, and other surfaces must be strong enough to withstand the impact of direct animal contact. Configuration and treatment of contact surfaces must minimize or eliminate protrusions, changes in elevation, and sharp corners to minimize bruising and injuries and to improve the efficiency of cattle handling. Proper lighting permits inspection of animals in feedlots and other cattle housing systems and provides safer working conditions for animal care personnel. Maintenance of facilities should be timely and ongoing.

FEED AND WATER

Diets for beef cattle should be formulated according to the recommendations of the NRC. Formulation of diets should consider factors such as environmental conditions, breed or biological type, sex, and production demands for growth, gestation, or lactation. Feed and water should be offered to cattle in ways that minimize contamination by urine, feces, and other materials. Feed bunks should be monitored daily and contaminants or spoiled feed should be removed. In most situations, feed should be available at all times.

However, restricted feeding of high-energy diets may be practiced to meet maintenance requirements or targeted levels of production. When restricted feeding is practiced, feed must be uniformly distributed in the bunk to allow all cattle to have simultaneous access to the diet. When high-energy diets are fed, increased attentiveness should be given to possible occurrence of diet-related health problems such as grain overload, lactic acidosis, and bloat.

Abrupt changes in diets should be avoided. Feed deprivation for more than 24 h should be avoided, and feed deprivation for any length of time must be justified in the animal use protocol. Cattle can vary considerably in body weight and condition during the course of grazing and reproductive cycles. Feeding programmes should allow animals to regain the body weight that is lost during the normal periods of negative energy balance.

Confined cattle should have continuous free access to a source of water, except before surgery or weighing if the research or animal care protocol requires such restriction. When continuous access to water is not possible, water should be available ad libitum at least once daily and more often if hot weather conditions exist or cattle have high levels of metabolizable energy intake for purposes of achieving high output.

Under winter range conditions, Degen and Young found that snow can be used as a water source for beef cows and growing calves. However, there was evidence that the snow resulted in reduced water intakes as evidenced by compensatory water intake when water was reintroduced following 84 d of consuming water in the form of snow. When snow was the only source of water, total water intake reductions averaged approximately 10 per cent among the cattle groups. The quantity and, possibly, quality of water available will influence animal comfort, especially under hot conditions. Evaporation of moisture from the skin surface or respiratory tract is the primary mechanism used by the animals to lose excess body heat in a hot environment. Estimates of daily water requirements for beef cattle are reported in NRC.

During summer months, in particular, waterer space available and water intake per animal becomes extremely important. Under these conditions, Mader found that as much as 3 times the normal waterer space may be needed to allow for sufficient room for all animals to access and benefit from available water. Additional waterer space recommendations are provided by MWPS.

ADEQUATE CARE OF CATTLE AND CALVES IN HUSBANDRY

Adequate care of cattle and calves is especially important for establishing and maintaining optimal immune system function. Good husbandry can minimize health problems and infectious diseases. The risk of disease and mortality in young calves is related to immune status. It is critical that newborn calves nurse or ingest colostrum soon after birth. The health of young growing cattle should be assessed regularly pre-and postweaning. Animal care personnel should be taught to recognize signs of illness and external parasites.

Alert caretakers should have the ability to perceive appropriate behaviour and posture. A system of monitoring calves through critical stress periods such as weaning should be established. Any sick or injured calves should be treated promptly. Daily records should be kept. For cattle reared in close confinement assessments should be done at least once daily and more often if cattle have been stressed or potentially exposed to conditions in which their health could be compromised. In general, confined feedlot cattle, especially new incoming cattle, require more frequent observations than non-confined cattle because of the greater probability of animal health being compromised due to comingling, dehydration, digestive problems, respiratory problems, and interaction of any of these factors with environmental stress. Signs of healthy calves are alert ears and clear eyes, no signs of diarrhea, and, upon arising, resumption of a normal standing posture after stretching. For feedlot cattle provided energy-dense diets, caretaker knowledge of acidosis and management regimens necessary to minimize digestive problems are essential. Appropriate medication and vaccination programmes should be used to reduce the incidence of disease and mortality, improve cattle health and performance, and ensure that no illegal residues occur in the carcass. Treatment and vaccination schemes should be based on veterinary advice and experience.

Weaning

In typical beef cow/calf production systems, calves are artificially weaned from their dams by physical separation. This process, albeit important to the efficiency of the cowherd, can be stressful to both the cow and calf. The most common weaning procedure involves an abrupt separation of cows and calves resulting in increased walking and vocalization and decreased eating and resting. An alternative to abrupt weaning and permanent separation is a period of fenceline contact between cows and calves in adjacent but separate pastures.

This weaning management alternative has been shown to decrease vocalization and walking and increase the time spent resting and grazing. This fenceline weaning procedure may also decrease the incidence of calf illness. Within the weaning pasture or pen, a mature cow can be included in the group of freshly weaned calves. This "trainer" cow can assist in introducing the weaned calves to the location and facilitating consumption of feed and water.

Despite the weaning process selected, it is important that weaned calves be provided access to clean water and a source of feed and/or forage. To encourage intake, highly palatable forage and feed sources are recommended until calves become accustomed to the separation from their dams. Additionally, feed and water sources should be placed close to the perimeter of the fenceline, because calves will typically spend a majority of their time in these areas as they seek to reunite with their dams.

Social Environment

Cattle are social animals. Each individual in the group should have sufficient access to the resources necessary for comfort, adequate well-being, and optimal performance. Mixing, crowding, group composition, and competition for limited resources are part of the social environment and in some circumstances, may be social stressors for certain cattle. Generally, cows from similar environments but from different social groups can be mixed with little or no long-term adverse effect on performance; however, because introduced cows may be the recipients of aggression, the number of mixing episodes should be minimized. Mixing of older cattle, especially bulls, results in more fighting than occurs when younger cattle are mixed. Fighting and mounting can be a problem associated with keeping bulls in social groups and can present a significant welfare problem if not managed carefully. Attempts should be made to keep bulls in stable social groups and to minimize mixing. When feed, water, or other resources critical for comfort or survival are limited, or when large differences exist among cattle in size or other traits related to position in the social order, some animals may be able to prevent others from gaining access to resources.

In properly designed facilities, all individuals should have sufficient access to feed, water, and resting sites to minimize the correlation between position in the social order and productive performance. Proper animal care includes observation of groups and of individuals within groups to ensure that each individual has adequate access to the resources necessary for optimal comfort, welfare, and performance.

Floor or Ground Area

Area recommendations for open lots and barns are listed in Table. Every animal should have sufficient space to move about at will, adequate access to feed and water, a comfortable resting site, and the opportunity to remain reasonably dry and clean. These suggested recommendations alone do not ensure that an ideal environment exists; however, in some cases these conditions can be met with less than the recommended area.

The area required is affected by type and slope of floor or soil surface, amount of rainfall, amount of sunshine, season, group size, and method of feeding. Open feedlot pens need to be sloped to promote drainage away from feed bunks, waterers, pen dividers, and resting areas. Space allocations are related directly to slope. In temperate Midwestern climates, the following relationships have been found to be workable: 2 per cent slope or less: 37 to 74 m^2 per animal; 2 to 4 per cent slope: 23 to 37 m^2; and 4 per cent or greater slope: 14 to 23 m^2.

Space allocations can be less in drier regions of the country. In the Southwest, at 0 per cent slope, typical allocations are 14 to 23 m^2 per animal. In other regions, space allocations may need to be increased above Midwestern norms in consideration of such factors as soil type and rainfall distribution. The area requirements for cattle are greatly influenced by group size. One animal housed separately in a pen requires the greatest amount of floor area on a per-animal basis. As group size increases, the amount of area required per individual decreases.

When an animal is housed individually, the minimum pen width and length should be at least equal to the length of the animal from nose tip to tail head when the animal is standing in a normal erect posture. Acceptable indoor pen floor surfaces for beef cattle include unfinished concrete, grooved concrete, concrete slats, expanded metal, plastic-covered metal flooring, and rubberized mat. The floor surface in stanchions and metabolism stalls may be concrete, expanded metal, wood, rubberized mat, or a combination of materials that provides support for the animals' bodies; does not damage hooves, feet, legs, and tails; and can be cleaned.

STANDARD AGRICULTURAL PRACTICES

For beef cattle, management procedures may be performed by properly trained, non-professional personnel. These include, but are not limited to, vaccinating, dehorning and castrating young cattle, horn-tipping, ear-tagging, branding, weighing, implanting, use of hydraulic and manual chutes for restraint, roping, hooftrimming, routine calving assistance, ultrasound pregnancy checking, feeding, and watering.

Other husbandry and health practices used in beef cattle research and teaching that similarly may be performed by properly trained, non-professional personnel, but that require special technical training and advanced skill levels, include artificial insemination, electroejaculation, pregnancy palpation, embryo flushing and transfer, non-routine calving assistance and dystocia treatment, emergency cesarean section, retained placenta treatment, and dehorning and castration of older cattle. One of the main animal husbandry concerns is that of pain and distress, especially pain inflicted from standard husbandry procedures. Dehorning, castration, and branding are husbandry procedures that can cause pain and discomfort; nevertheless, these procedures are justified as a management tool to minimize injuries or other problems associated with confining horned cattle and commingling bulls.

Dystocia Management

Matings should be planned to lessen the genetic probability of dystocia. When dystocia does occur, proper care and assistance at calving can decrease injury or death of both calves and heifers/cows. Parturition without complication is common in beef cows. Therefore, before administering assistance to a cow experiencing difficulty with calving, personnel should be familiar with the stages associated with approaching parturition and the signs of normal delivery.

As a general rule, females should be examined within 30 to 60 min following presentation of feet, nose, or fetal membranes if delivery of the calf does not appear imminent. However, heifers or cows exhibiting signs of a malpresentation, oversized fetus, fetal anomaly, or other obvious complication must be assisted immediately. Facilities should be provided that are designed for restraint of cows and heifers experiencing dystocia.

Because many animals, especially heifers, lie down during the obstetrical procedure, sufficient space should be provided to permit adequate freedom of movement. It is important that the obstetrical restraint facility be fitted with side gates, both of which are hinged at the head end, so that the animal can become fully recumbent and the obstetrical procedure can be performed with safety and efficiency. In dystocia cases where fetal presentation appears to be compromised or there appears to be a disparity between the size of the fetus and the diameter of the birth canal, assistance of the delivery by personnel appropriately trained in the judicious use of a fetal extractor may be attempted. In general, if more than slight traction is required on the fetal extractor, the procedure should be stopped and a veterinarian called immediately to perform a caesarean section or fetotomy. Use of excessive force can damage the calf and/or dam and lead to suffering and/or death. Strict sanitation should be used with all obstetrical procedures.

Vaccinations and Drug Administration

Vaccinations are a key component to any herd health programme. Care should be taken to ensure the proper use, handling, and storage of vaccines and approved or investigational drugs. The preferred site of injection is the neck for either intramuscular or subcutaneous injections; however, for investigational drugs used in research, alternate sites of administration may be required or preferred as dictated by the research protocol. Investigators and animal care staff should utilize best management practices associated with the use of syringes and handling needles. Use and regular replacement of disposable syringes and needles is highly recommended to avoid excessive trauma and disease transmission.

Castration

Castration of male beef cattle is performed to reduce aggressiveness, prevent physical danger to other animals in the herd and to handlers, enhance reproductive control, manage genetic selection, and satisfy consumer preferences regarding taste and tenderness of meat. Castration of young bulls is a necessary management practice in beef production.

Several methods for castrating cattle are acceptable, including surgical removal of the testicles using a knife or scalpel to open the scrotum and cutting or crushing the spermatic cords with an emasculatome or emasculator. Bloodless procedures utilizing specialized rubber rings or surgical tubing bands are available to create devitalization and eventual sloughing of the tissues below the ring or band.

High-tension banding systems may be used with appropriate veterinary supervision and/or training in those situations where surgical castration may predispose to postsurgical complications or when surgical castration is not appropriate because of its effect on research protocol. The castration method used should take into account the animal's age and weight, the skill level of the technician, environmental conditions, and facilities available as well as human and animal safety.

Whatever the method of castration, the procedures should be conducted by, or under the supervision of, a qualified, experienced person and carried out according to castration equipment manufacturer recommendations and accepted husbandry practices.

Surgical castration is normally a short-term event with short-term duration of pain-associated responses. Bloodless castration has been associated with lower short-term pain indicators but longer chronic pain indicators. Bloodless castration should be used when surgical castration may predispose to postsurgical complications or when surgical castration is not appropriate because of its effect on the research protocol.

Castration is least stressful when performed at or shortly after birth, but lower stress is reported if performed before 2 or 3 months of age or before animals reach a body weight of 230 kg. It is strongly recommended that calves be castrated at the earliest age possible. It may be desirable to inject local anesthetic in the scrotum of calves heavier than 230 kg when surgical methods of castration are used or when the spermatic cords are crushed.

Topical local anesthetics may also be used on open wounds. Improved animal performance, as one potential indicator of improved animal welfare, has not been observed in animals locally anesthetized at the time of castration. It should be recognized that the effect of anesthetic agents is short-lived. Nevertheless, procedures should be implemented to minimize pain and discomfort, especially in older cattle.

Castration of older, heavier bulls should be performed only by skilled individuals. When it is necessary to castrate these heavier bulls, techniques and procedures to control bleeding must also be used. No advantage to use of anesthesia is apparent when bloodless castration is practiced. The possibility of infection should be given additional consideration after castration. Equipment should be sterilized, and facilities should be clean and sanitized.

Infection following castration can be minimized by keeping the animals in a clean area and away from excessive mud or contaminants following the procedure until the wound is healed. If tetanus is a common disease associated with the premises, or if a bloodless castration method is utilized, the herd health veterinarian should schedule a prophylactic tetanus immunization programme.

Dehorning

Horns on cattle can cause bruises and other injury to other animals, especially during transport and handling. Horns on adult cattle also can be a hazard to humans. Hornless cattle require less space in the feedlot and at the feed bunk. Polled breeds should be used whenever possible. Disbudding and dehorning of cattle in the United States is not currently regulated.

The Canadian Veterinary Medical Association recommends that disbudding be performed within the first week of life. In the United Kingdom, disbudding with a hot iron is preferred to dehorning and it is advised that this should be performed before cattle reach the age of 2 mo. In Australia, dehorning without local anesthesia or analgesia is restricted to animals less than 6 mo old.

Calves suffer less pain and stress, have less risk of infection, and have better growth rates when dehorning is performed at a very young age. Stafford and Mellor found that the use of local anesthetics virtually eliminated the escape behaviour of calves associated with the dehorning process and that a 2-h delay was observed in the cortisol response to horn amputation. Whenever possible, the use of a local anesthetic is encouraged when dehorning.

When horned breeds of cattle are selected, dehorning should be performed under the supervision of experienced persons using proper techniques. The horn buds should be removed at birth or within the first month after birth by several means, including hot cauterizing irons, cauterizing chemicals, a sharp knife, or commercially available mechanical devices. It is strongly recommended that calves be dehorned at the earliest age possible. When it is necessary to remove horns from older cattle, methods that minimize pain and bleeding and prevent infection should be employed.

Dehorning should be performed by a person knowledgeable and experienced in the appropriate procedures. Appropriate restraint and local anesthesia to control pain should be used when cattle older than 1 mo of age are dehorned. Cattle should be monitored for hemorrhage and infection following dehorning.

Adult cattle should be dehorned if aggressive behaviour is displayed towards herd mates or humans. Dehorning may temporarily depress the growth of cattle. In the event that bunk and pen space are ample, then tipping the horn may be considered as an alternative to minimize potential bruising or injury of pen mates. However, Ramsay reported that, after transport, carcass bruises were as common among tipped cattle as among horned ones.

Identification Methods

Proper animal identification is essential to research, facilitates record keeping, and aids in the routine observation and repeat identification of cattle. Methods of identification include skin colour markings, ear tagging, tattooing, hot branding, freeze branding, and electronic identification. Ear tags are best used in conjunction with a more permanent form of identification such as a tattoo or brand, as ear tags are sometimes lost.

Hot branding the hide is utilized as a means of identification; however, loss in hide value and studies indicating that freeze branding is less painful than hot branding have begun to minimize the use of hot branding. Alternatives to hot branding should be considered. However, skin and hair colour in addition to a limited access to liquid nitrogen or dry ice in extensive range operations may affect the ability to achieve a quality freeze brand.

At some locations, branding is required by law. Both hot branding and freeze branding should be performed by trained personnel to minimize skin contact with the branding device to only that required to achieve a useful brand. Advent of a national animal identification system (NAIS) in the form of visual or *radio frequency identification* (RFID) ear tags serve as an additional means of identification. As this system will become standard for all cattle as part of a national programme, managers of beef cattle as part of resident herds used in research should comply with the established guidelines.

Implanting

Implanting of cattle is a management practice for the administration of growth promotants and potentially as a means of delivery of investigational compounds used in research. For proper absorption and maximum response, implants should be placed correctly and in the correct location. Traditionally, implants are placed beneath the skin on the back side of the middle third of the ear; however, alternate implantation sites may be required as designated by the research protocol. Proper disinfection of the implant site is required to prevent infection. Care should be taken not to injure major blood vessels or the cartilage of the ear when implanting in the ear location. Utilization of best management practices associated with the use of the implant device and correct needle-handling procedures are required by suitably trained personnel.

SPECIAL CONSIDERATIONS AND INTENSIVE LABORATORY FACILITIES

Some research and teaching situations require that beef cattle be housed under intensive laboratory conditions. Cattle may be kept in metabolism stalls, stanchions, respiration chambers, or environmental chambers. Housing cattle in such facilities should be avoided unless required by the experimental protocol and then should be for the minimum amount of time necessary to accomplish the teaching or research objective. Cattle that are held or penned temporarily in crowded areas, frequently disturbed, or come into close contact with humans, or exposed to unfamiliar conditions or laboratory/teaching settings should have calm dispositions and be adapted to frequent contact with animal care personnel and to those conditions that could result in the animal having an adverse reaction.

In some cases, it may be advantageous to train such animals to a halter. Time spent preparing cattle for use in a laboratory improves the quality of research and the safety of both the animals and the humans. Cattle should not be housed in isolation unless approved by the Animal Care and Use Committee for specific experimental requirements. Whenever possible, cattle should be able to maintain visual contact with others. Unless the experimental protocol has special requirements for lighting, all animal rooms should be designed to minimize variation in light intensity. During light periods, the minimum light intensity for intensively housed cattle is 70 lx. If possible, a diurnal light-dark cycle should be used and a standard daily schedule established. Excreta should be removed from enclosed laboratories at least once daily. Pens or stalls should be washed thoroughly at the beginning of every trial.

If excreta or other foreign materials such as wasted feed cannot be adequately removed through daily cleaning, additional washing may be needed during a trial. The method of collection of feces and urine from cattle in metabolism stalls, stanchions, and chambers depends on the design and construction of the unit. Additional management may be needed to keep animals clean when they are housed in stalls or stanchions. Cattle may need to be washed and curried regularly to maintain cleanliness and to avoid fly infestations. Pens, stalls, and stanchions should be large enough to allow cattle to stand up or lie down without difficulty and should be long enough to allow cattle to maintain a normal standing position. Because of the operating costs associated with singlepass ventilation systems in controlled environmental facilities, partial recirculation of exhaust air from animal rooms is common and acceptable in many studies. In facilities designed to recirculate even a small part of the exhausted air, treatment is necessary to remove odorous compounds, gases, and particulate matter. Cattle maintained in some laboratory environments have their activity restricted more than cattle in production settings. The length of time that cattle may remain in stanchions, metabolism stalls, or environmental chambers before removal to a pen or outside lot for additional exercise should be no longer than that necessary for conducting the study.

Opportunities for regular exercise should be considered if they do not disrupt the experimental protocol; care must be taken in moving animals from the laboratory to the outside environment for exercise when a large temperature differential exists. If cattle are to be housed in such laboratory environments for more than 3 wk then particular attention should be given to alertness of the animal; appetite; fecal and urinary outputs; and condition of the feet, legs, and hock joints. Rubber mats or suitable alternatives should be used to increase the comfort of cattle maintained for lengthy periods on hard surfaces.

Use of Beef Cattle in Biomedical Research

Relative size, cost of maintaining beef cattle, and the use of alternate animal models in biomedical research have largely minimized the use of beef cattle in this regard. Nevertheless, beef cattle have played a role in understanding such maladies as lysosomal storage diseases and hemochromatosis, among others, which have similarities to diseases, often genetically based, found in humans; therefore beef cattle may serve as highly valuable biomedical models in some cases. In addition, the potential use of cattle as bioreactors for the production of human gene products or pharmaceuticals in milk, blood, urine, or tissues may further extend the use of beef cattle for biomedical applications. Standards for the care and welfare of beef cattle used in biomedical research should be the same as that applied to all beef cattle.

However, institutional or biomedical funding agencies may require more specific disease entry testing requirements for cattle used in biomedical research, in addition to having more stringent procedures with respect to adherence to alternate oversight committee guidelines for reporting, housing, observation and care procedures than might be utilized or generally accepted under typical agricultural research and production systems. In some cases, in which in vitro reproductive technologies are used for the production of beef cattle in research, maturation, fertilization, manipulation, and/or culture, differences can exist in fetal morphology, physiology, and in the expression of developmentally important genes that may require alteration in management strategy. For example, cattle produced in this manner may exhibit "large calf syndrome" and therefore may require extra assistance at calving. The animal biotechnology sector continues to grow, with significant advancements being made that may directly and indirectly affect beef cattle research, and it is important to recognize that alterations through the genetic engineering of beef cattle may similarly require alterations in beef cattle care practices. With respect to genetic engineering, unanticipated results from genetic modifications have been observed in several genetically engineered species that require diligence on the part of the researcher and animal care staff in assessing animal welfare. However, the general standards of care associated with genetically engineered and cloned beef cattle should be the same as that applied to all beef cattle unless the specific genetic modification requires an alteration in management within the research environment to specifically facilitate animal welfare.

EUTHANASIA

The USDA and Food Safety and Inspection Service (FSIS) *Humane Slaughter of Livestock* regulations, floors of livestock pens, ramps, and driveways of harvest facilities shall be constructed and maintained so as to provide good footing for livestock. Animals shall have access to water in all holding pens and, if held longer than 24 h, access to feed. Also, for animals held overnight there shall be sufficient room in the holding pens for the animals to lie down.

The AVMA *Guidelines on Euthanasia* lists several methods of euthanasia that are appropriate for ruminants. Intravenous administration of barbiturates, potassium chloride used in conjunction with general anesthesia, and penetrating captive bolt are acceptable means of euthanasia in all cases. Other conditionally acceptable methods include intravenous administration of chloral hydrate, gunshot to the head, and electrocution.

In all cases, euthanasia should only be performed by trained individuals. Agents that result in tissue residues cannot be used for the euthanasia of ruminants intended for human or animal food, unless those agents are approved by the Food and Drug Administration. Carbon dioxide is the only chemical currently used in euthanasia of food animals that does not lead to tissue residues.

Use of carbon dioxide is generally not recommended for euthanasia of larger animals. The carcasses of animals euthanized by barbiturates may contain potentially harmful residues, and such carcasses should be disposed of in a manner that prevents them from being consumed by humans or animals. Dying, diseased, and disabled livestock shall be provided with a covered pen sufficient to protect them from adverse climatic conditions. Incurably ill or injured animals in chronic pain or distress should be humanely euthanized as soon as they are diagnosed as such and according to AVMA recommended procedures. Their disposal should be accomplished promptly by a commercial rendering service or other means according to applicable ordinances and regulations.

Bibliography

A. Hemantaranjan: *Advancements in Iron Nutrition Research*, Scientific Research, 1995.

A. Poshadri and Aparna Kuna: *A Handbook of Food Techno's*, New India Publishing Agency, Delhi, 2013.

A.D. Dholakia: *Delicious Seafood Recipes*, Daya Publications, Delhi, 2010.

A.S. Bawa, P.S. Raju and O.P. Chauhan: *Food Science*, New India Publishing Agency, Delhi, 2013.

Alice Bradley: *Cooking for Profit; Catering and Food Service Management,* American School of Home Economics, Chicago, 1922.

Andrew L. Winton and Kate Barber Winton : *A Handbook of Structure and Composition of Foods*, Agrobios Publication, Delhi, 2013.

Arindam Ramaswamy: *Health and Food Science*, Oxford Book Company, Jaipur, 2010.

Arnold C. Long: *Advances in Seafood Biochemistry: Composition and Quality*, Cyber Tech Publications, Delhi, 2009.

Ashok K Chauhan and Ajit Varma: *Microbes : Health and Environment*, I K International, Delhi, 2007.

Avantina Sharma: *Text Book of Food Science and Technology*, Ibdc Publishers, Lucknow, 2010.

Bruno Dorin and Frederic Landy: *Agriculture and Food in India: A Half-Century Review, From Independence to Globalization*, Manohar Publications, Delhi, 2009.

C S Jain: *A Complete Book on Health and Nutrition*, Cyber Tech Publications, Delhi, 2009.

Dinesh Arora: *Biotech's Dictionary of Food Technology*, Biotech Books, Delhi, 2005.

Gaurav Gandhi: *Hotel Management Food and Food Services*, Random Publications, Delhi, 2012.

H.A. Modi: *Dairy Microbiology*, Aavishkar Publishers, Delhi, 2009.

H.A. Modi: *Fermented Milk Products*, Aavishkar Publication, Delhi, 2011.

J. O. Dahl, and J. H. Breland: *Food Standards Handbook for Quantity Cookery,* Whitlock Press, Stamford Conn, 1945.

J. O. Dahl: *200 Ways to Control Food Costs in Quantity Cookery,* The Dahls, Stamford Conn, 1941.

Jagmohan Negi and Gaurav Manoher: *Food and Beverage Management*, Himalaya Publications, Delhi, 2011.

Jagmohan Negi, M.J. Gaurav and Ritushka and Suniti: *Food And Beverage Service Skills Techniques Operation And Management*, Kanishka Publications, Delhi, 2011.

M Lakshmi Narasaiah: *Economic Growth and Food Security*, Discovery Publishing House, Delhi, 2008.

Madhulika Bhatnagar: *Encyclopaedia of Catering Technology, Food Service and Hospitality Management, Vol. I to III*, Anmol Publications, Delhi, 2008.

Margaret M. Walsh: *Food and Nutrition Manual for Institutions,* Welfare Federation of Cleveland, Cleveland, 1950.

Parminder K. Bhandari: *Achieving Nutritional Goals*, Sonali Publications, Delhi, 2010.

R.P. Saxena: *Food Services and Catering Management*, Centrum Press, Delhi, 2010.

Renu Singh: *Food and Beverage Management*, Delhi, Rajat Publications, 2012.

Roberta Angelone: *Cooking Italian Food in India*, Westland Publisher, 2013.

Udai Veer: *Elements of Food Science*, Anmol Publications, Delhi, 2007.

Umesh Prasad: *Food Science and Nutrition*, Sonali Publications, Delhi, 2011.

Urvashi Nandal: *A Handbook of Foods and Nutritional Biochemistry: A Complete Source Book*, Agrobios Publications, Delhi, 2013.

Vijaya Rao: *Armies Wars and Their Food*, Cambridge University Press, Delhi, 2012.

Index